AF615051

CELL ADHESION AND HUMAN DISEASE

The Ciba Foundation is an international scientific and educational charity (Registered Charity No. 313574). It was established in 1947 by the Swiss chemical and pharmaceutical company of CIBA Limited—now Ciba-Geigy Limited. The Foundation operates independently in London under English trust law.

The Ciba Foundation exists to promote international cooperation in biological, medical and chemical research. It organizes about eight international multidisciplinary symposia each year on topics that seem ready for discussion by a small group of research workers. The papers and discussions are published in the Ciba Foundation symposium series. The Foundation also holds many shorter meetings (not published), organized by the Foundation itself or by outside scientific organizations. The staff always welcome suggestions for future meetings.

The Foundation's house at 41 Portland Place, London W1N 4BN, provides facilities for meetings of all kinds. Its Media Resource Service supplies information to journalists on all scientific and technological topics. The library, open five days a week to any graduate in science or medicine, also provides information on scientific meetings throughout the world and answers general enquiries on biomedical and chemical subjects. Scientists from any part of the world may stay in the house during working visits to London.

Ciba Foundation Symposium 189

CELL ADHESION AND HUMAN DISEASE

1995

JOHN WILEY & SONS

Chichester · New York · Brisbane · Toronto · Singapore

Published in 1995 by John Wiley & Sons Ltd
Baffins Lane, Chichester
West Sussex PO19 1UD, England
Telephone National Chichester (0243) 779777
International (+44) (243) 779777

Ciba Foundation Symposium 189
ix+243 pages, 34 figures, 9 tables

Library of Congress Cataloging-in-Publication Data

Cell adhesion and human disease / editors, Joan Marsh and Jamie A. Goode.
p. cm.—(Ciba Foundation symposium; 189)
Includes bibliographical references and index.
ISBN 0-471-95279-6
1. Cell adhesion—Congresses. 2. Cell adhesion molecules—Congresses. 3. Physiology, Pathological—Congresses. I. Marsh, Joan. II. Goode, Jamie. III. Series.
RB113.C268 1995
616.07—dc20 94-39499
CIP

British Library Cataloguing in Publication Data

A catalogue record for this book is
available from the British Library

ISBN 0 471 95279 6

Phototypeset by Dobbie Typesetting Limited, Tavistock, Devon.
Printed and bound in Great Britain by Biddles Ltd, Guildford.

Contents

Symposium on Cell adhesion and human disease, held at the Ciba Foundation, London 17–19 May 1994

Editors: Joan Marsh (Organizer) and Jamie A. Goode

Participants

J. N. W. N. Barker St John's Institute of Dermatology, St Thomas's Hospital, Lambeth Palace Road, London SE1 7EW, UK

W. Birchmeier Max-Delbrück-Centrum für Molekulare Medizin, Robert-Rössle-Strasse 10, D-13125 Berlin, Germany

M. J. Elices Cytel Corporation, 3525 John Hopkins Court, San Diego, CA 92121, USA

B. Ernst Zentrale Forschungslaboratorien, Ciba-Geigy Ag, Rosenthal R1060.3.34, CH-4002 Basle, Switzerland

A. Etzioni Department of Pediatrics, Rambam Medical Center, B. Rappaport Medical School, IL-31096 Haifa, Israel

D. R. Garrod CRC Epithelial Morphogenesis Research Group, School of Biological Sciences, University of Manchester, 3.239 Stopford Building, Oxford Road, Manchester M13 9PT, UK

D. O. Haskard Department of Medicine (Rheumatology Unit), Royal Postgraduate Medical School, Hammersmith Hospital, Du Cane Road, London W12 0NN, UK

P. Herrlich Kernforschungszentrum Karlsruhe Gmbh, Institut für Genetik, PO Box 3640, D-76021 Karlsruhe, Germany

N. Hogg Imperial Cancer Research Fund, PO Box No 123, Lincoln's Inn Fields, London WC2A 3PX, UK

M. J. Humphries School of Biological Sciences, University of Manchester, 2.205 Stopford Building, Oxford Road, Manchester M13 9PT, UK

R. O. Hynes *(Chairman)* Center for Cancer Research, Massachusetts Institute of Technology, 77 Massachusetts Avenue, Cambridge, MA 02139-4307, USA

G. Koopman *(Ciba Foundation Bursar)* Department of Pathology, Academic Medical Center, University of Amsterdam, Meibergdreef 9, NL-1105 AZ Amsterdam, The Netherlands

M. Labow Department of Biotechnology, Hoffman La Roche, 340 Kingland Street, Nutley, NJ 07110-1199, USA

S. V. Ley University Chemical Laboratory, University of Cambridge, Lensfield Road, Cambridge CB2 1EW, UK

S. T. Pals Department of Pathology, Academic Medical Center, University of Amsterdam, Meibergdreef 9, NL-1105 AZ Amsterdam, The Netherlands

R. B. Parekh Oxford GlycoSystems Ltd, Unit 4, Hitching Court, Blacklands Way, Abingdon OX14 1RG, UK

J. S. Pober Boyer Center for Molecular Medicine, Yale University School of Medicine, POB 9812, 295 Congress Avenue, New Haven, CT 06536-0812, USA

G. Riethmüller Institut für Immunologie, Universität München, Goethestrasse 31, D-80336 München, Germany

R. Rothlein Department of Immunology, Boehringer Ingelheim Pharmaceutical Inc, 900 Ridgebury Road, PO Box 368, Ridgefield, CT 06877, USA

Z. M. Ruggeri Department of Molecular and Experimental Medicine, Scripps Research Institute, 10666 North Torrey Pines Road, La Jolla, CA 92037, USA

F. Sánchez-Madrid Seccion de Immunologia, Universidad Autónoma de Madrid, Hospital de la Princesa, Calle de Diego de Leon 62, E-28006 Madrid, Spain

S. Shaltiel Department of Chemical Immunology, The Weizmann Institute of Science, PO Box 26, IL-76 100 Rehovot, Israel

A. Sonnenberg Division of Cell Biology, Netherlands Cancer Institute, Plesmanlaan 121, NL-1066 CX Amsterdam, The Netherlands

J. R. Stanley Dermatology Branch, National Cancer Institute, Building 10, Room 12N238, Bethesda, MD 20892, USA

P. Verrando Unité INSERM 387 Adhésion Cellulaire, Hôpital de Sainte-Marguerite, 270 Boulevard de Sainte Marguerite, F-13277 Marseille Cedex 9, France

D. D. Wagner Center for Blood Research, Harvard Medical School, 800 Huntington Avenue, Boston, MA 02115, USA

R. K. Winn Department of Surgery, University of Washington School of Medicine, Harborview Medical Center, 325 Ninth Avenue, ZA-16, Seattle, WA 98104-2499, USA

Chairman's introduction

Richard O. Hynes

Howard Hughes Medical Institute and Center for Cancer Research, Massachusetts Institute of Technology, Cambridge, MA 02139-4037, USA

Cell adhesion plays an important role in many disease states. These include cancer (invasion and metastasis), thrombosis, inflammatory diseases and problems arising from ischaemia/reperfusion injury (heart attacks, stroke, organ transplantation, frostbite). In all these situations cells show alterations in their adhesive properties; basically, they stick where they should not or they stick too much.

In recent years, advances in our understanding of the molecular basis of cell adhesion have revealed the existence of families of cell surface receptors. Each cell adhesion event involves one, or more often several, adhesion receptors from the various families of molecules. Detailed analyses of these receptors have in many cases defined their binding sites. This presents the opportunity to block the binding sites using antibodies, peptides, carbohydrate groups or synthetic analogues of these reagents. In this way, it is hoped that undesired adhesive elements can be blocked, providing novel therapies for human diseases.

The challenge is to define with adequate specificity which of the many receptors are involved in a given disease and which present the best targets for therapeutic intervention. Potential anti-adhesive therapeutic drugs can enter human clinical trials only after adequate testing in animal systems. In this context, recent advances in methods for generating mice with alterations in specific genes allow the development of animal models of human genetic deficiencies and also model systems lacking one or more adhesion receptors. In this way, the roles of individual adhesion systems in specific diseases can be defined precisely, allowing better planning of therapeutic approaches.

Finally, there are some questions I think we should attempt to answer during this symposium (and this is obviously not an exhaustive list). First, which adhesion molecules should we try to block the function of in any given disease and how will we find out which are the best ones to target? Second, once we have identified them, which strategy should we choose for blocking the function of these molecules? Third, is there the possibility of gene therapy for some adhesion diseases? The final question is a serious issue, which I hope we can discuss: how does one go about blocking adhesion in a chronic fashion in the treatment of diseases such as rheumatoid arthritis, psoriasis and cancer?

P-selectin knockout: a mouse model for various human diseases

Denisa D. Wagner

The Center for Blood Research, Harvard Medical School, 800 Huntington Avenue, Boston, MA 02115, USA

Abstract. P-selectin is a transmembrane adhesion receptor specific to platelets and endothelial cells. It has an N-terminal lectin domain that recognizes specific carbohydrate moieties on monocytes, neutrophils and some other subsets of leukocytes. P-selectin is stored in granules and is expressed on the plasma membrane only after the cells are stimulated by vascular injury or during inflammation. Physiologically P-selectin is likely to be involved in the recruitment of leukocytes that promote wound healing and fight infection. There are many disorders in which the excessive recruitment of leukocytes is characteristic, including chronic inflammation, atherosclerosis, arthritis, diabetes, asthma and reperfusion injury. Because certain cancer cells also express the ligand for P-selectin it is possible that this receptor is involved in metastasis. To study the specific role of P-selectin in these pathological processes, we have prepared a mouse lacking P-selectin through gene targeting. Leukocyte interaction with the vessel wall is defective in these animals as leukocytes do not roll in the mesenteric venules and their extravasation at sites of inflammation and vessel injury is limited. We are testing these animals in models of the various diseases mentioned above in order to evaluate when the absence of P-selectin is beneficial.

1995 Cell adhesion and human disease. Wiley, Chichester (Ciba Foundation Symposium 189) p 2–16

The vessel wall is no longer considered to be only an inert barrier to blood. For example, the endothelial cells forming the inside wall of vessels can modify the composition of their plasma membranes in response to environmental cues. This can happen either slowly, by *de novo* synthesis of new membrane components, or rapidly, by the mobilization of preformed membrane components to the plasma membrane. The endothelial cells can also modify their secreted products depending on the circumstances, thus influencing the environment. The general research interest of my laboratory is the rapid response of endothelium to injury. We have found that endothelial cells store adhesive proteins in storage granules called Weibel–Palade bodies (Weibel & Palade 1964), which are rapidly secreted at the time of a vascular injury. These proteins are von Willebrand factor, the soluble adhesion molecule for platelets (Wagner et al 1982) and P-selectin, the transmembrane receptor for leukocytes (Bonfanti et al 1989, McEver et al 1989). The granules exocytose their contents within minutes of endothelial stimulation with secretagogues such as histamine, thrombin, fibrin or complement components C5b–9 (Wagner 1993). Von Willebrand factor and P-selectin are also found together in platelet α-granules, from where they are released upon platelet activation. In this chapter I will discuss primarily P-selectin, as von Willebrand factor is covered elsewhere in this volume (Ruggeri 1995).

P-selectin is a member of the selectin family of adhesion receptors (Lasky & Rosen 1992, Bevilacqua & Nelson 1993). The name 'selectin', thought of by Bevilacqua (Bevilacqua et al 1991), indicates the function of the protein in selective interactions mediated by a lectin domain. Indeed, the Ca^{2+}-dependent (C-type) lectin domains of these molecules bind to specific negatively charged carbohydrate structures presented by various mucins (Springer 1994). The N-terminal lectin domain is followed by an epidermal growth factor-like domain and several complement-binding protein-like repeats. The proteins contain a single transmembrane domain and a short cytoplasmic tail. L-selectin, which is present on leukocytes, is also called the homing receptor, as it mediates homing of lymphocytes to the peripheral lymph nodes (Gallatin et al 1983). L-selectin is also present on neutrophils where it contributes to neutrophil adhesion and extravasation at sites of inflammation. E-selectin is specific to endothelial cells, but in contrast to P-selectin, it is synthesized only after exposure of the endothelium to inflammatory cytokines (Bevilacqua et al 1987), at which time it becomes directly expressed on the plasma membrane (Fig. 1). After internalization, E-selectin is rapidly degraded in the lysosomes. P-selectin, which is also endocytosed after its expression on the surface, is directed to the Golgi region where it is incorporated into nascent Weibel–Palade bodies (Fig. 1, Subramaniam et al 1993). P-selectin can therefore be involved both in acute processes, when it becomes available immediately after stimulation, and in chronic phenomena, when it may cycle several times between the membrane and the storage granule under repeated stimulation. In addition, the synthesis of P-selectin is up-regulated when cells are treated with endotoxin or inflammatory cytokines (Sanders et al 1992). Since the two endothelial selectins mediate primarily adhesion to neutrophils and monocytes (Bevilacqua & Nelson 1993) and may be present simultaneously on activated endothelial cells, their functions could overlap. To determine the specific functions of P-selectin, in both normal physiology and pathological conditions, we prepared P-selectin-deficient animals in collaboration with Richard Hynes. This was accomplished (Mayadas et al 1993) through homologous recombination in embryonic stem cells (Capecchi 1989). The engineered animals, which do not contain any P-selectin in their platelets and endothelial cells, are grossly normal and fertile.

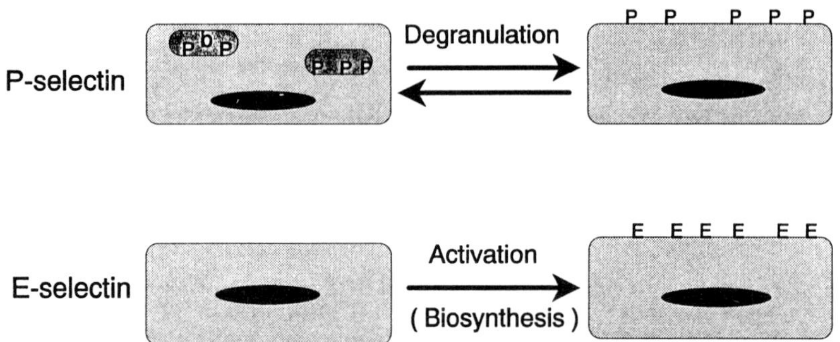

FIG. 1. The expression of P- and E-selectins on endothelial cells. P-selectin is stored in Weibel–Palade bodies and it becomes rapidly expressed on the plasma membrane upon degranulation. After endocytosis, the P-selectin returns to storage granules (Subramaniam et al 1993). In the presence of inflammatory cytokines endothelial cells begin to synthesize E-selectin. The expression may last many hours and eventually all the protein is degraded in lysosomes.

This shows that P-selectin is dispensible for normal embryonic development and angiogenesis. The disruption of the P-selectin gene did not affect the expression of E- and L-selectin, whose genes are located in close proximity to P-selectin (Watson et al 1990).

The total peripheral leukocyte and platelet counts in the P-selectin-deficient animals are similar to those of wild-type animals. In contrast, the basal neutrophil counts in the homozygous-deficient animals were two to three times higher than in the wild-type animals. Since the numbers of progenitors in the bone marrow were determined to be comparable, this difference is very likely due to a longer half-life of the neutrophils in the mutants. Indeed, by injecting radiolabelled human neutrophils into the tail vein of mutant and wild-type animals, Robert Johnson (unpublished results) showed that these cells survive longer in the bloodstream of the mutant mice.

Several lines of investigations have implicated the selectins in the first step in leukocyte extravasation, leukocyte rolling. First, P-selectin embedded in a lipid bilayer supports leukocyte rolling under physiological flow conditions *in vitro* (Lawrence & Springer 1991). Second, infusion of anti-L-selectin antibody significantly inhibits leukocyte rolling *in vivo* (von Andrian et al 1991, Ley et al 1991). To investigate the effect of the absence of P-selectin on leukocyte rolling, we performed intravital microscopy of the mouse mesentery. In this model (Atherton & Born 1973), the mesentery of anaesthetized animals is pulled out and spread onto a microscope stage, and the behaviour of leukocytes is directly observed and recorded. To increase rapidly the expression of P-selectin on the vessel wall, we treated the mesentery with the Ca^{2+} ionophore A23187, which we knew to be an excellent *in vitro* secretagogue for Weibel–Palade bodies

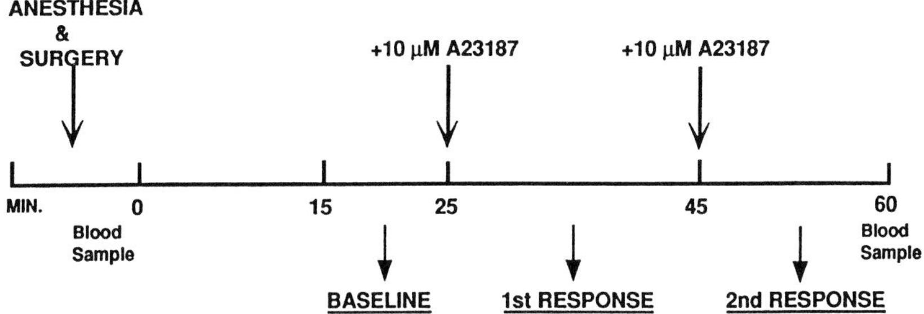

FIG. 2. Schematic representation of the experimental design of an intravital microscopy study. Following the 10 min baseline period, 10 μM calcium ionophore A23187 was applied to the mesentery by superfusion, at the times indicated. In the wild-type animals the baseline rolling was 10.5 ± 2.5, first response to the ionophore was 24.9 ± 7.1 and second response 17.3 ± 3.3 (values represent mean number of rolling leukocytes passing through a perpendicular plane per min, $\pm$SEM [$n = 8$ for wild-type, 5 for mutant]). In the P-selectin-deficient animals we did not observe any rolling leukocytes at baseline. First response was 0.10 ± 0.06 and second response 0.04 ± 0.04 (Mayadas et al 1993).

(Sporn et al 1986). The complete experimental design is presented in Fig. 2. In the wild-type mice, the numbers of leukocytes observed rolling on the vessel wall more than doubled after the ionophore treatment, as if the vessel were stickier. In contrast, there were no rolling leukocytes in the P-selectin-deficient mesentery under the baseline conditions and only one leukocyte for every 10 min after the ionophore treatment (Mayadas et al 1993). It is clear from these results that P-selectin plays a crucial role in the initial interaction of the leukocytes with the vessel wall. The L-selectin of the leukocyte cannot alone support baseline rolling. We have performed intravital microscopy studies using animals with experimentally induced peritonitis. Our preliminary results indicate that when the mesentery is inflamed (i.e. several hours after injection of thioglycollate), leukocytes can roll in the absence of P-selectin (Robert Johnson, unpublished observations). Therefore, under inflammatory conditions, other adhesion molecules can mediate leukocyte rolling. Interestingly, in the P-selectin-deficient mice after thioglycollate injection, the leukocytes roll much more slowly than in wild-type animals. This is likely due to a different strength of adhesive interactions in this case from that generated by P-selectin and its ligand. In addition, in the wild-type animals many more leukocytes are in contact with the vessel wall, demonstrating the importance of P-selectin even under the inflammatory conditions.

The defect in the initial contact of the leukocyte with the vessel wall in the P-selectin-deficient animals is likely the cause of an observed delay in extravasation (Mayadas et al 1993). Injection of thioglycollate in the peritoneum of wild-type animals stimulates a rapid onset of neutrophil extravasation, whereas in the P-selectin-deficient animals there is a 2 h delay. After this time, it is possible that a different molecule, perhaps E-selectin, may be expressed on the endothelium, thus allowing leukocyte extravasation. Currently, we are preparing mice deficient in both P- and E-selectin: it will be interesting to see whether the neutrophil extravasation to the peritoneum in response to thioglycollate will be ablated completely in these animals. We have also examined the effect of P-selectin deficiency on the recruitment of macrophages into chronically inflamed peritoneum. We have found that the numbers of macrophages recruited 48 h after a thioglycollate injection are significantly reduced in the absence of P-selectin (Robert Johnson, unpublished observation). This result further indicates that P-selectin can play a role in chronic inflammation.

Another chronic situation we have investigated is contact hypersensitivity. In the murine model, although the response is initiated by $CD4^+$ lymphocytes, the neutrophilic response is more intense than in humans. To elicit the contact hypersensitivity response, we sensitized mice with oxazolone, followed by injection of $[^{125}I]$iododeoxyuridine, which labels monocytes and lymphocytes. Twenty-four hours post-challenge, the mutant mice had 50% lower counts in sensitized ears than the wild-type mice. In addition, histological sections of the ears showed much less neutrophil infiltration in the mutant mice (M. Subramaniam, unpublished observations). Lack of P-selectin therefore affects the recruitment of both mononuclear cells and neutrophils in this model.

Weibel–Palade bodies and α-granules are massively exocytosed at the time of a vascular injury, when thrombin is generated (Fig. 3). The released von Willebrand factor is important for the formation of a platelet plug and the P-selectin probably participates in the recruitment of phagocytic cells to fight infection and to clear debris. To evaluate the actual role of P-selectin in the recruitment of phagocytes to the site of a wound, we have adapted the following wound healing model: full thickness skin excisional wounds (3–4 mm in diameter) were generated on the flanks of wild-type and P-selectin-deficient mice. Biopsies were taken from killed mice at intervals of hours or days following wounding. Staining of the 1 h wound sections showed neutrophils exiting along post-capillary venules and veins in the wild-type mice, but very few in the mutant mice. By 4 h, the numbers of extravasated neutrophils in the mutant mice approached those in the wild-type animal wounds (M. Subramaniam, unpublished observations). It will be interesting to see whether the P-selectin deficiency has an effect on wound closure.

To summarize, we have learned the following about the phenotype of the P-selectin-deficient mice: the mice have an elevated neutrophil count that is probably due to defective neutrophil clearance; their leukocytes do not roll in mesenteric venules in the absence of an inflammatory stimulus; their neutrophils show a 2 h lag in extravasation at sites of inflammation and at wound sites; macrophage recruitment to chronically inflamed peritoneum is reduced; and the mice have a significantly reduced contact hypersensitivity response.

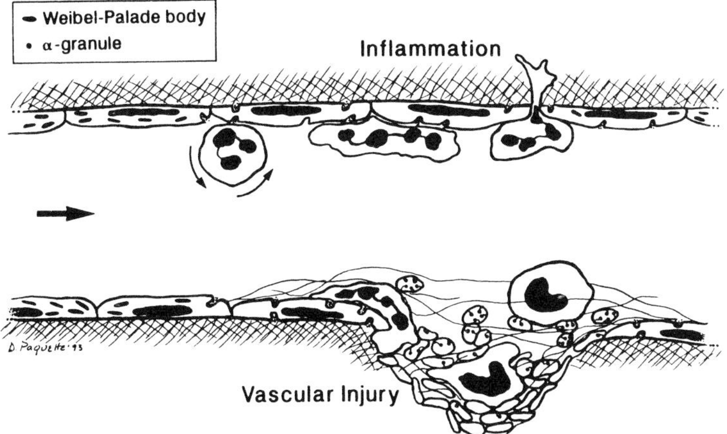

FIG. 3. Diagram of an injured venule. The degranulation of platelet α-granules and endothelial Weibel–Palade bodies mobilizes P-selectin to the cell surface where it serves as receptor for monocytes, neutrophils and other subsets of leukocytes. It mediates leukocyte rolling on endothelium at sites of inflammation and recruitment of phagocytes to a wound. (Reproduced with permission from Wagner 1993.)

Because of the striking phenotype of the P-selectin-deficient mice, these animals could provide an interesting model system with which to study the role of P-selectin in various diseases and pathological conditions. The most obvious condition in which P-selectin is likely to play a role is inflammation. P-selectin binds myeloid cells and, more recently, has been shown also to bind natural killer cells and subpopulations of memory T cells of both the $CD4^+$ and $CD8^+$ type (Moore & Thompson 1992, Damle et al 1992, de Bruijne-Admiraal et al 1992), and may therefore mediate migration of all these cell types to inflammatory sites. Some rather well known diseases are linked to inflammation. For example, in insulin-dependent diabetes mellitus, the defect in insulin secretion results from the destruction of pancreatic islets by a chronic, slowly evolving, inflammatory process called insulitis, which is thought to be of autoimmune origin. Histological studies show islet infiltration with macrophages and both $CD4^+$ and $CD8^+$ T cells (Rossini et al 1993). It would be interesting to know whether P-selectin is one of the adhesion molecules involved in this inflammatory process.

Atherosclerosis is another prominent disease in which P-selectin is very likely to play a role. In this vascular disorder, monocytes adhere to injured endothelium (that is likely to express P-selectin), extravasate and become engorged with lipid vesicles while remaining under the endothelium (Ross 1986). Some of the lipid may be obtained from phagocytosed platelets (Sevitt 1986), known to be present in advanced lesions, a process that could also be mediated by P-selectin. Mice fed a high-fat diet develop atherosclerotic lesions and therefore can be used as animal models for atherosclerosis (Paigen et al 1985). Because genetic manipulation of mice is possible, the role of an individual gene in the complex process of atherosclerosis can now be studied. This should be done for P-selectin.

Selectins were also shown to bind to carbohydrate on some cancer cells and therefore may be involved in the metastatic spread of cancer. E-selectin has been shown to bind to colon cancers (Rice & Bevilacqua 1989, Lauri et al 1991) and we have shown that P-selectin mediates binding of platelets to neuroblastoma and small-cell lung cancer (Stone & Wagner 1993). Since P-selectin, and not E-selectin, is expressed on platelets in addition to endothelial cells, it may have a unique importance for the development of successful metastases. Experimental animal models of metastasis have convincingly demonstrated the involvement of platelets in this process (Gasic et al 1968, 1973). With the existence of the P-selectin-deficient mouse one will actually be able to test the hypothesis that this selectin participates in the haematogenous spread of cancer cells.

These are just a few examples of diseases and conditions where P-selectin may be implicated. Understanding of the situations where adhesive interactions through P-selectin do play a role will undoubtedly lead to the development of new drugs that inhibit these interactions and therefore arrest or reverse the disease process. We hope that the development of the P-selectin-deficient animal model will bring us closer to this goal.

Acknowledgements

This work was supported by National Institutes of Health grants HL41002 and HL53756.

References

Atherton A, Born GVR 1973 Relationship between the velocity of rolling granulocytes and that of the blood flow in venules. J Physiol 233:157–165

Bevilacqua MP, Nelson RM 1993 Selectins. J Clin Invest 91:379–387

Bevilacqua MP, Pober JS, Mendrick DL, Cotran RS, Gimbrone MA 1987 Identification of an inducible endothelial–leukocyte adhesion molecule. Proc Natl Acad Sci USA 84:9238–9242

Bevilacqua E, Butcher B, Furie B et al 1991 Selectins: a family of adhesion receptors. Cell 67:233

Bonfanti R, Furie BC, Furie B, Wagner DD 1989 PADGEM (GMP-140) is a component of Weibel–Palade bodies of human endothelial cells. Blood 73:1109–1112

Capecchi MR 1989 Altering the genome by homologous recombination. Science 244:1288–1292

Damle NK, Klussman K, Dietsch MT, Mohaghephpour N, Aruffo A 1992 GMP-140 (P-selectin/CD62) binds to chronically stimulated but not resting $CD4^+$ T lymphocytes and regulates their production of proinflammatory cytokines. Eur J Immunol 22:1789–1793

de Bruijne-Admiraal LG, Modderman PW, Von dem Borne AE, Sonnenberg A 1992 P-selectin mediates Ca(2+)-dependent adhesion of activated platelets to many different types of leukocytes: detection by flow cytometry. Blood 80:134–142

Gallatin WM, Weissman IL, Butcher EC 1983 A cell-surface molecule involved in organ-specific homing of lymphocytes. Nature 304:30–34
Gasic GJ, Gasic T, Stewart C 1968 Antimetastatic effects associated with platelet reduction. Proc Natl Acad Sci USA 61:46–52
Gasic GJ, Gasic TB, Galanti N, Johnson T, Murphy S 1973 Platlet–tumor-cell interactions in mice. The role of platelets in the spread of malignant disease. Int J Cancer 11:704–718
Lasky LA, Rosen SD 1992 Carbohydrate-binding adhesion molecules of the immune system. In: Gallin JI, Goldsten IJ, Snyderman R (eds) Inflammation: basic principles and clinical correlates. Raven Press, New York, p 1–13
Lauri D, Needham L, Martin-Padura I, Dejana E 1991 Tumor cell adhesion to endothelial cells: endothelial leukocyte adhesion molecule-1 as an inducible adhesive receptor specific for colon carcinoma cells. J Natl Cancer Inst 83:1321–1324
Lawrence MB, Springer TA 1991 Leukocytes roll on a selectin at physiologic flow rates: distinction from and prerequisite for adhesion through integrins. Cell 65:859–873
Ley K, Gaehtgens P, Fennie C, Singer MS, Lasky LA, Rosen SD 1991 Lectin-like cell adhesion molecule 1 mediates leukocyte rolling in mesenteric venules in vivo. Blood 77:2553–2555
McEver RP, Beckstead JH, Moore KL, Marshall-Carlson L, Bainton DF 1989 GMP-140, a platelet α-granule membrane protein, is also synthesized by vascular endothelial cells and is localized in Weibel–Palade bodies. J Clin Invest 84:92–99
Mayadas TN, Johnson RC, Rayburn H, Hynes RO, Wagner DD 1993 Leukocyte rolling and extravasation are severely compromised in P-selectin-deficient mice. Cell 74:541–554
Moore KL, Thompson LF 1992 P-selectin (CD62) binds to subpopulations of human T lymphocytes and natural killer cells. Biochem Biophys Res Commun 186:173–181
Paigen B, Marrow A, Brandon C, Mitchell D, Holmes PA 1985 Variation in susceptibility to atherosclerosis among inbred strains of mice. Atherosclerosis 57:65–73
Rice GE, Bevilacqua MP 1989 An inducible endothelial cell surface glycoprotein mediates melanoma adhesion. Science 246:1303–1306
Ross R 1986 The pathogenesis of atherosclerosis—an update. N Engl J Med 314:488–500
Rossini AA, Greiner DL, Friedman HP, Mordes JP 1993 Immunopathogenesis of diabetes mellitus. Diabetes Rev 1:43
Ruggeri ZM 1995 Von Willebrand's disease and the mechanisms of platelet function. In: Ccll adhcsion and human disease. Wiley, Chichester (Ciba Found Symp 189) p 35–50
Sanders WE, Wilson RW, Ballantyne CM, Beaudet AL 1992 Molecular cloning and analysis of in vivo expression of murine P-selectin. Blood 80:795–800
Sevitt S 1986 Platelets and foam cells in the evolution of atherosclerosis. Histological and immunological studies of human lesions. Atherosclerosis 61:107–115
Sporn LA, Marder VJ, Wagner DD 1986 Inducible secretion of large biologically potent von Willebrand factor multimers. Cell 46:185–190
Springer TA 1994 Traffic signals for lymphocyte recirculation and leukocyte emigration: the multistep paradigm. Cell 76:301–314
Stone JP, Wagner DD 1993 P-selectin mediates adhesion of platelets to neuroblastoma and small cell lung cancer. J Clin Invest 92:804–813
Subramaniam M, Koedam JA, Wagner DD 1993 Divergent fates of P- and E-selectins after their expression on the plasma membrane. Mol Biol Cell 4:791–801
von Andrian UH, Chambers JD, McEvoy LM, Bargatze RF, Arfors KE, Butcher EC 1991 Two-step model of leukocyte–endothelial cell interaction in inflammation: distinct roles for LECAM-1 and the leukocyte beta 2 integrins in vivo. Proc Natl Acad Sci USA 88:7538–7542

Wagner DD 1993 Weibel–Palade body: the storage granule for von Willebrand factor and P-selectin. Thromb Haemostasis 70:105–110
Wagner DD, Olmsted JB, Marder VJ 1982 Immunolocalization of von Willebrand protein in Weibel–Palade bodies of human endothelial cells. J Cell Biol 95:355–360
Watson ML, Kingsmore SF, Johnston GI et al 1990 Genomic organization of the selectin family of leukocyte adhesion molecules on human and mouse chromosome 1. J Exp Med 172:263–272
Weibel ER, Palade GE 1964 New cytoplasmic components in arterial endothelia. J Cell Biol 23:101–112

DISCUSSION

Labow: In the thioglycollate model of inflammation, where are the neutrophils that invade the peritoneum coming from?

Wagner: The leukocytes are coming from the blood vessels that are in contact with the peritoneum, because they can sense the cytokines that are produced in a response to the thioglycollate, such as IL-8.

Ruggeri: I was intrigued by your comments about the interaction that you see in the P-selectin knockout between the neutrophils and the vessel wall during inflammation. You suggested that another adhesion mechanism is responsible for the ability of these cells still to interact and crawl on the surface. Why didn't you also see that in the wild-type? This would be a mechanism that should exist in the wild-type. Is it possible that it has been overlooked in the wild-type because only a minority of cells use this pathway? If there is such a pathway, why does it work only in a small number of cells?

Wagner: We may be looking here just at small subsets of leukocytes that can roll slowly, for example, on E-selectin. This may also be the case in wild-type mice, but it seems that the majority of the cells are rolling fast on P-selectin. Another possibility is that P-selectin, being a larger molecule than E-selectin, may be sticking out further into the bloodstream. Therefore, P-selectin may grab the leukocyte first and shift it to another P-selectin molecule—like playing volleyball. So the leukocyte would be flying from one P-selectin to another and E-selectin could never reach it.

Ruggeri: If it is true that there is a molecule that mediates a stronger adhesion than that mediated by P-selectin, at some point you should be able to differentiate the relative functional role of the different pathways by increasing the flow rate. On the arterial surface you don't see any rolling, so there's obviously a shear rate above which selectins cannot function. Consequently, if you set up a model where you can vary the flow under controlled conditions, you may be able to clarify what the different mechanisms could be.

Wagner: Again, you're absolutely right. In the arterioles the shear rate could be so high that although the selectins are expressed, they are no longer able to function properly. We have compared blood vessels of very similar location and diameter in two sets of animals that were treated in the same way. We would expect the shear rates in the two to be similar. It would be interesting to vary shear rates, but at the moment we don't know how to do it. Lawrence & Springer (1993) have immobilized P-selectin and E-selectin on plates and studied the leukocytes rolling under flow conditions in this system. They saw that the leukocytes rolled more slowly on E-selectin, which suggests that this may be the molecule expressed in our knockout mice under inflammatory conditions.

Ernst: Under *in vitro* conditions, P-selectin shows up immediately after stimulation and is gone after 20–30 min, whereas E-selectin is expressed only after 2–4 h. What is the picture *in vivo*?

Wagner: Our studies with cultured endothelial cells have shown that P-selectin is not destroyed after surface expression, but it returns to the Weibel–Palade bodies. Therefore it could become available again. We hypothesize that this is how P-selectin could also be involved in chronic situations *in vivo*.

Haskard: You're talking about E-selectin as being a possible molecule on which the leukocytes could roll in the absence of P-selectin, but what about L-selectin? Have you thought of doing experiments with anti-L-selectin antibodies to see if that abrogates the residual rolling?

Wagner: Others have done this with wild-type animals. Injection of anti-L-selectin antibodies inhibits leukocyte rolling, but not completely (Ley et al 1991, von Andrian et al 1991). Maybe the L-selectin on the leukocyte has to find a ligand—perhaps CD34—on the endothelium, and at the same time P-selectin has to find a carbohydrate on the leukocyte. It could be that for optimal rolling, independent interactions mediated through two selectins are needed.

Hynes: There's not much of a depression in the 'baseline' leukocyte rolling in the L-selectin knockout that Tom Tedder has made (unpublished results).

Wagner: That is correct. P-selectin is clearly the most important selectin for rolling under these baseline conditions.

Some people have also proposed that P- and L-selectin may be 'holding hands', i.e. that P-selectin may be recognizing carbohydrate presented by L-selectin, because L-selectin contains sialyl Lewis X structures. But that is not always the case. For example, with Reina Mebius we have looked at adhesion of wild-type leukocytes to high endothelial venules in peripheral lymph nodes of P-selectin knockout mice, an interaction known absolutely to require L-selectin. But these knockout lymph nodes didn't have P-selectin. So, if L- and P-selectin had to 'hold hands', then the leukocytes wouldn't have been able to bind to the P-selectin-deficient lymph nodes, but they do bind just as well as to wild-type lymph nodes.

Shaltiel: Has anyone made an estimate of the number of points of contact during 'good' rolling? How would rolling be affected by changing the density of P-selectin expression on the cell? For instance, if you were to do an experiment in the presence of increasing levels of antibodies against P-selectin, how would this affect rolling?

Wagner: Tim Springer's lab has some *in vitro* results (Lawrence & Springer 1993) showing that the density of a selectin affects the speed of leukocyte rolling: the velocity of rolling decreases with increase in selectin density. The speed of rolling is also dependent on shear stress. But you reach a plateau for the density of a particular selectin when more shear no longer increases velocity.

Shaltiel: So the level of selectin release would form a sort of steady-state—would this be a mechanism of regulating the process?

Wagner: Yes, you could regulate or fine-tune the rolling process by controlling the release of P-selectin from the Weibel–Palade bodies, because secretion from Weibel–Palade bodies (as we know from our own experience) is not an all-or-none phenomenon—you can have partial secretion of the storage granules. So you could up-regulate or down-regulate the rolling or change the speed of rolling in that way. In addition, you could control rolling by up-regulating *de novo* synthesis of selectins by inflammatory cytokines.

Birchmeier: You mentioned cancer cells: have you actually initiated studies to look at metastasis, for instance, by using wild-type cells in mutant mice?

Wagner: We are starting to do these studies.

Pober: Part of the general paradigm for the difference between P- and E-selectin is that P- selectin pre-exists in resting endothelial cells, but that's clearly not always the case. You raise the issue of the role of P-selectin in atherosclerosis; it's not so clear that large arteries have much P-selectin in their Weibel–Palade bodies. What is known about regulation of P-selectin expression? How do you set the baseline?

Wagner: No one has really studied P-selectin expression in different blood vessels. We have seen some P-selectin expression in arteries, others have also reported this (McEver et al 1989). There is always the possibility that P-selectin, like E-selectin, is up-regulated by *de novo* synthesis, because cytokines and TNF have been shown to up-regulate P-selectin (Sanders et al 1992, Weller et al 1992). It is also possible that different levels of stimulation are necessary to bring the P-selectin to the cell surface in an arteriole than are needed in a venule, because we never see leukocyte rolling in arterioles. But this might be because the slight inflammation we created in our experiments is not enough to cause secretion from the Weibel–Palade bodies in arterioles.

Ruggeri: This is an interesting question; it might be related to the fact that von Willebrand factor expression in arterial cells is very heterogeneous. It's widely assumed that von Willebrand factor is a marker for endothelial cells. Although this is certainly true *in vitro* (every cultured endothelial cell makes von Willebrand factor), when you look *in vivo*, at the level of both the protein and mRNA, you see a tremendous heterogeneity. There are actually some endothelial structures and vessels that lack von Willebrand factor entirely, particularly in arteries. A careful study of the expression of von Willebrand factor would be extremely valuable.

Hynes: If you take Denisa Wagner's theory that P-selectin is recruiting monocytes in atherosclerosis, isn't it possible that the P-selectin is coming from the platelets, not necessarily just from the endothelial cells?

Pober: That depends on whether or not there are ligands on the vessel wall to interact with P-selectin. You could have a platelet bridge between a leukocyte and the endothelial cell surface, but I'm not aware that there's any clear evidence that endothelial cells, for example, have P-selectin ligands.

Hynes: I agree, but what if the platelets got stuck there by some other process? They could then express P-selectin on activation.

Wagner: For example, if they were bound to von Willebrand factor deposited there as a result of some injury.

Pober: The evidence from the lipid-fed primates in Russell Ross' serial morphological studies does not implicate platelets in the adhesion of monocytes to the foam cell lesions (reviewed in Ross 1993). The initial recruitment of monocytes appears to be platelet independent.

I think everyone has partially ignored the issue of whether or not the regulation of expression of P-selectin is controlled by the same mechanisms as E-selectin, and whether one could think about pharmacological targeting to inhibit synthesis by the same approaches that are being widely tested for inhibiting synthesis of E-selectin.

Wagner: This is regulation at the mRNA level, but for P-selectin, in addition, one could try to target the regulated secretory pathway. Inhibitors of regulated secretion exist, such as microtubule-depolymerizing agents, but one would have to use something more subtle to inhibit the secretion of Weibel–Palade bodies. An inhibitor of Weibel–Palade body secretion might also be a useful therapeutic agent for people with thrombotic problems, which may be a result of release of the large von Willebrand factor multimers stored in these granules.

Sonnenberg: Does anybody know what the function of P-selectin is on platelets?

Wagner: In collaboration with Bruce and Barbara Furie, we showed a while ago that P-selectin on platelets mediates adhesion to monocytes and neutrophils (Larsen et al 1989). This interaction has now been confirmed to be mediated solely by P-selectin because, for the platelets in P-selectin knockout mice, this leukocyte binding is totally ablated. But we don't yet know whether or not this is the only function of P-selectin on platelets. When the platelets have been isolated and washed, we don't see a defect in platelet aggregation in response to collagen, for example. But *in vivo* this aggregation doesn't happen with just platelets as there are always leukocytes present and the cells 'talk' to each other. We don't know whether a thrombus that is formed *in vivo* in the P-selectin-deficient mice is the same size and is held together with the same strength as one in the wild-type mice. We are now looking for a machine that would allow us to measure platelet aggregation in whole blood, and it is not easy to come by.

Ruggeri: Aggregation is the most unrealistic scenario for platelet function, because it involves platelets in suspension sticking to one another—that's not how things work *in vivo*. If you don't have a surface and if things don't start on a surface, you are looking at an artefact.

Hynes: It's nice to hear a platelet person say that!

Ruggeri: Studying aggregation is probably like studying transgenic mice: it tells you a lot but it doesn't tell you the full story and you always have to go back to real life.

Etzioni: In the wound healing experiment, you showed that after 4 h there's an almost normal number of neutrophils in the area, whereas in the contact hypersensitivity model, even after 8 h, there was a marked increase in the neutrophil count. Do you have an explanation for this?

Wagner: Perhaps the stimulus is different in the hypersensitivity reaction and the other adhesion molecules are expressed more slowly.

Etzioni: Do you think that there is no E-selectin expression and that it is just a P-selectin-dependent reaction?

Wagner: I don't know. We will have to get some E-selectin knockout mice and look at their contact hypersensitivity response.

Labow: Have you followed the time course of the expression of E- and P-selectin by immunohistochemistry in these mice?

Wagner: No. One reason this is difficult is that antibodies that recognize mouse antigens are scarce. We plan to use some of your reagents to look at E-selectin. We intend to make P- and E-selectin double knockouts, to see what will happen in the absence of both endothelial selectins. Selectins may not be the whole story—in some regions the blood flow is so slow that you may not need selectins at all and other adhesion molecules may mediate leukocyte rolling.

Hynes: It's becoming clear that every situation has to be looked at separately. We can't generalize that it's P-selectin first and then E-selectin (it often isn't) and we can't generalize about how long P-selectin stays around (sometimes it may be transient and at other times it clearly isn't). We're just going to have to look at each situation separately.

Barker: Concerning your allergic contact dermatitis model: the theory is that you apply your allergen and it's taken up by antigen-presenting cells in the epidermis (Langerhans' cells), which migrate to lymph nodes. It is known that Langerhans' cells express certain selectin ligands on their cell surface which are up-regulated, at least in human skin (Ross et al 1994). Are the P-selectin-deficient mice more difficult to sensitize? This deficiency may well inhibit Langerhans' cell trafficking.

Wagner: We really don't know—there may be a defect in the way these mice are sensitized as well.

Barker: You mentioned that the neutrophil response in the skin of these mice is intense after induction of contact allergy; in humans you don't see any neutrophils at all—it's a completely lymphocytic infiltrate.

Wagner: The response in mice is very different from humans in that it's much more neutrophilic, although it is initiated by T lymphocytes.

Pober: There are new data from Phil Askenase (unpublished results) which indicate that in this kind of contact sensitivity model the magnitude of the response is dependent on platelets, which are presumably delivering serotonin and other vasoactive mediators to the reaction. The defect in P-selectin-deficient animals that you're seeing, rather than being involved in the ultimate recruitment of the neutrophils into the lesion through P- versus E-selectin adhesion, may instead reflect a platelet defect in terms of delivering the vasoactive mediators. Also, it's a fairly controversial question as to whether T cells utilize P-selectin for homing to inflammatory sites, and your methods are not directly measuring T cell infiltration *per se*. Nevertheless, I suspect you will see fewer T cells in the lesions, because T cell recruitment is often dependent upon the initial infiltration of other leukocytes. Although human reactions are almost entirely mononuclear at 24 h (when they are usually measured), at 4 h they are predominantly neutrophilic—the human neutrophil response subsides very quickly, whereas the murine one doesn't.

Wagner: We are now quantitating the difference in the monocyte infiltration and also labelling the $CD4^+$ T cells in the dermis, so we should be able to see the effect of P-selectin deficiency on the recruitment of the different classes of leukocytes.

Garrod: Presumably, you keep the P-selectin knockout mice in a controlled clean environment, but are there any differences in their survival or their responses to infection and injury compared with normal mice?

Wagner: They are not kept under any special conditions; we don't keep them like immunodeficient mice, for instance. Sometimes we even keep them in the lab for several days and they do fine. They don't seem to be especially susceptible to infection, but we haven't yet done a controlled study with an infectious agent. I expect if we really challenged these mice, they would have problems.

Garrod: Presumably they get small injuries?

Wagner: They may bite each other and they rub their noses on the cages, but we haven't noticed much of a difference between the P-selectin knockouts and wild-type mice. Their longevity and fertility are similar.

References

Larsen E, Celi A, Gilbert GE et al 1989 PADGEM protein: a receptor that mediates the interaction of activated platelets with neutrophils and monocytes. Cell 59:305–312

Lawrence MB, Springer TA 1993 Neutrophils roll on E-selectin. J Immunol 151:6338–6346

Ley K, Gaehtgens P, Fennie C, Singer MS, Lasky LA, Rosen SD 1991 Lectin-like cell adhesion molecule 1 mediates leukocyte rolling in mesenteric venules *in vivo*. Blood 77:2553–2555

McEver RP, Beckstead JH, Moore KL, Marshall-Carlson L, Bainton DF 1989 GMP-140, a platelet α-granule membrane protein, is also synthesized by vascular endothelial cells and is localized in Weibel–Palade bodies. J Clin Invest 84:92–99

Ross R 1993 The pathogenesis of atherosclerosis: a perspective for the 1990s. Nature 362:801–809

Ross EL, Barker JNWN, Allen MH, Chu AC, Groves RW, MacDonald DM 1994 Langerhans' cell expression of the selectin ligand Sialyl Lewis X. Immunology 81:303–308

Sanders WE, Wilson RW, Ballantyne CM, Beaudet AL 1992 Molecular cloning and analysis of *in vivo* expression of murine P-selectin. Blood 80:795–800

von Andrian UH, Chambers JD, McEvoy LM, Bargatze RF, Arfors K-E, Butcher EC 1991 Two-step model of leukocyte–endothelial cell interaction in inflammation: distinct roles for LECAM-1 and the leukocyte β_2 integrins *in vivo*. Proc Natl Acad Sci USA 88:7538–7542

Weller A, Isenmann S, Vestweber D 1992 Cloning of the mouse endothelial selectins. J Biol Chem 267:15716–15183

Creation and characterization of E-selectin- and VCAM-1-deficient mice

Lia Kwee*, Daniel K. Burns†, John M. Rumberger†, Chris Norton*, Barry Wolitzky*, Robert Terry*, Kathleen M. Lombard-Gillooly*, David J. Shuster*, Frank Kontgen‡, Colin Stewart°, Kim McIntyre*, Scott Baldwin§ and Mark A. Labow*

**Roche Research Center, Department of Biotechnology and °Roche Institute of Molecular Biology, Hoffman-La Roche Inc., 340 Kingland Street, Nutley, NJ 07110-1199, †Glaxo Research Institute, Research Triangle Park, NC 27709, USA, ‡WEHI, The Royal Melbourne Hospital, Victoria, Australia and §Wistar Institute, 3601 Spruce Street, Philadelphia, PA 19104-4268, USA*

Abstract. A variety of adhesion molecules have been identified which mediate the interaction of leukocytes with endothelial cells. In order to define the role of individual molecules in inflammation we have produced lines of mice which are deficient in the synthesis of specific adhesion molecules. Null mutations were introduced into the genes encoding E-selectin or vascular cell adhesion molecule-1 (VCAM-1) in embryonic stem cells and these cells were used to produce lines of mice carrying the mutation. E-selectin-deficient mice were viable and exhibited no developmental defects. The roles of E- and P-selectin in the influx of neutrophils were examined using these mice. The data suggest that the two selectins are functionally redundant in mediating neutrophil emigration in a model of chemically induced peritonitis. VCAM-1-deficient mice are not viable. Analysis of VCAM-1 gene expression in wild-type embryos and phenotypic analysis of VCAM-1 −/− embryos suggests that VCAM-1 is required for development of the extraembryonic circulatory system and the embryonic heart.

1995 Cell adhesion and human disease. Wiley, Chichester (Ciba Foundation Symposium 189) p 17–34

The mechanism by which leukocytes traffic through blood vessels involves at least three distinct steps (for review see Springer 1994). The initial interaction under normal conditions of flow results in the rolling of leukocytes along the endothelium. The second step appears to be the arrest and firm adhesion of the leukocyte to the vessel wall. Finally, leukocytes undergo diapedesis. The selectin family of proteins is thought to mediate the rolling of leukocytes while the integrins expressed on leukocytes and immunoglobulin superfamily member proteins (Ig proteins) expressed on endothelial cells are thought to mediate firm adhesion and diapedesis. Three different selectins have been identified (see Bevilacqua 1993 for review). E-selectin and P-selectin are expressed on activated endothelial cells while a third, L-selectin, is expressed on most circulating leukocytes. One Ig protein of particular interest is vascular cell adhesion molecule-1 (VCAM-1). VCAM-1 is expressed on activated endothelial cells and mediates their binding to a number of leukocytes positive for the integrin $\alpha_4\beta_1$ (also known as VLA-4).

In order to begin to define the roles of specific adhesion molecules in inflammatory processes, we have created lines of mice deficient in expression of specific adhesion molecules. Mice deficient in expression of E-selectin and VCAM-1 have been developed and in this paper we describe their initial characterization.

Characterization of E-selectin-deficient mice

The gene for E-selectin is organized in a manner similar to those of other selectins as shown in Fig. 1 (Becker-Andre et al 1992). All selectins contain an N-terminal lectin-like (Lec) domain which is critical for binding of the carbohydrate component of the E-selectin ligand(s). This domain is encoded on a separate exon as are the epidermal growth factor-like domain and domains containing homology to complement regulatory proteins (CR domains). Because the Lec domain is critical for E-selectin function, a targeting vector was produced which introduces a large insertion and frameshift mutation (an MC1*neo* cassette; Thomas & Capecchi 1987) into the Lec-encoding exon of the E-selectin locus. A detailed description of the construction of this vector will be published elsewhere. The targeting vector was used to transform the W9.5 embryonic stem (ES) cell line and ES cell culture was carried out as previously described (Stewart et al 1992, Abbondanzo et al 1993). Southern blot analysis of DNA from *neo*^{r} ES cell clones identified a large number of clones containing a disrupted E-selectin allele as evidenced by the detection of a unique 6 kb XbaI restriction fragment. Blastocyst injection of two of the targeted clones produced germline-transmitting chimeric mice from each clone. Inter-cross experiments between heterozygous animals resulted in the production of mice homozygous for the E-selectin mutation at Mendelian frequencies (data not shown). The $-/-$ animals were of normal size and weight and both sexes were fully fertile. Thus, the E-selectin gene does not appear to be required for normal mouse development.

In order to determine if the E-selectin mutation was truly a null mutation we carried out a variety of experiments. Initially, a reverse-transcriptase polymerase chain reaction (RT-PCR) assay was carried out to determine if any normal E-selectin mRNA could be detected in the $-/-$ animals. E-selectin $-/-$ mice and wild-type litter mates were injected with IL-1α to induce expression of E-selectin and RNAs from several tissues were isolated and used to produce cDNA. Each cDNA clone was then analysed by PCR using

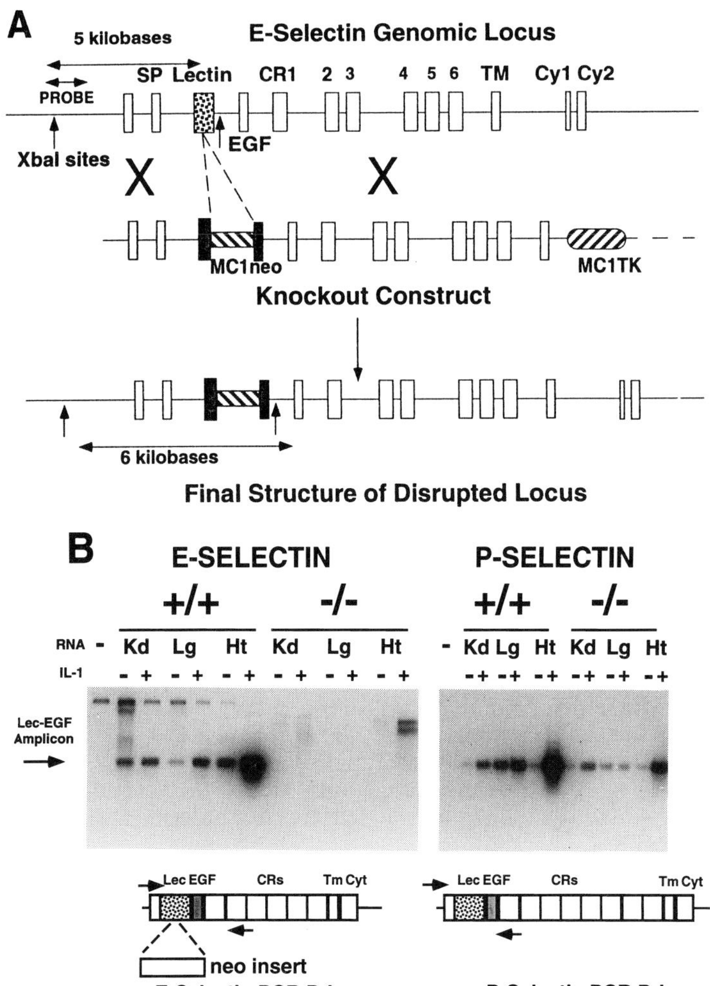

FIG. 1. Disruption of the murine E-selectin gene. (A) The genomic structure of the murine E-selectin gene is shown along with the structure of the knockout construct and the predicted structure of the targeted allele. Also shown are the restriction enzyme sites used to identify animals carrying the mutant allele. The identification of these animals will be described in detail elsewhere. The mutation introduced into the locus is an MC1*neo* cassette. An MC1*tk* gene was also constructed and inserted into the vector for negative selection. (B) Reverse-transcriptase polymerase chain reaction (RT-PCR) analysis of selectin message in wild-type and E-selectin $-/-$ mice. RNA from various tissues of mice injected with phosphate-buffered saline (PBS) or PBS and human interleukin (IL)-1α were subjected to RT-PCR using primers located within the indicated Lec and CR domains of the murine E-selectin and P-selectin cDNAs. PCR products were then analysed by Southern blotting and hybridization to specific Lec–EGF domain cDNA probes. EGF, epidermal growth factor; Kd, kidney; Lg, lung; Ht, heart; Tm, transmembrane; Cyt, cytoplasmic.

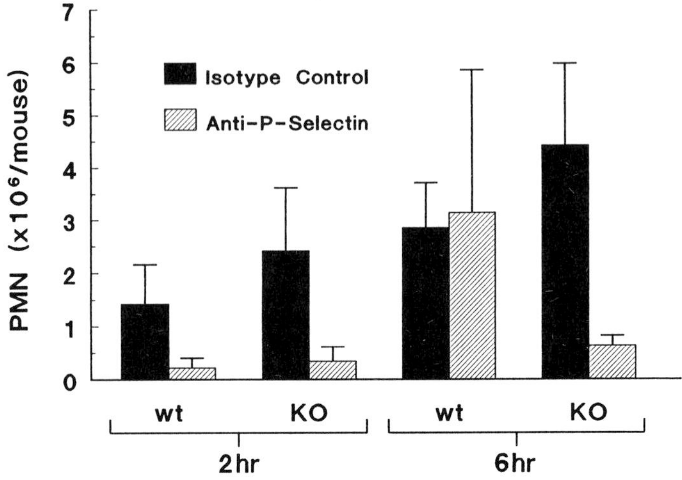

FIG. 2. The roles of E-selectin and P-selectin in neutrophil influx into the peritoneum after intraperitoneal injection of thioglycollate. Four mice in each group were injected with thioglycollate and the peritoneal lavage assayed for number of polymorphonuclear leukocytes (PMNs) at 2 h or 6 h after injection. Mice were treated with thioglycollate and either 5H1, an anti-murine P-selectin antibody, or an equal amount of an isotype control. (Error bars represent standard deviation, $n = 4$.) wt, wild-type; KO, E-selectin-knockout mice.

primers flanking the *neo* insertion. As a control, PCR reactions were also carried out with P-selectin primers. Whereas E-selectin amplicons of the expected size were detected in all tissues of the wild-type mouse, no normal sized amplicons were detected using cDNA from the −/− mouse. The only amplicons detected in the −/− mice were present in very low amounts and were approximately 1 kb larger than those produced by the normal E-selectin mRNA. These amplicons appear to represent E-selectin mRNA containing the entire *neo* insertion indicating that no functional mRNA was produced. In contrast, similar amounts of P-selectin mRNA were detected by the RT-PCR assay in both wild-type and −/− mice. Consistent with these results, no E-selectin protein was detected either by immunoprecipitation of *ex vivo* labelled protein or by *in situ* immunohistochemistry of heart and lung tissue of −/− mice (data not shown). Thus, the mutation introduced into the E-selectin locus was a true null mutation.

In order to test the role of E-selectin in an acute inflammatory response, we compared neutrophil influx in wild-type and E-selectin-deficient mice during chemically induced peritonitis. Figure 2 shows the number of neutrophils present in the peritoneum at various times after intraperitoneal injection of thioglycollate. As shown, both wild-type and −/− mice contained approximately equal numbers of neutrophils 2 h after injection. In addition, no difference was observed between wild-type and E-selectin-deficient mice 6 h after thioglycollate injection.

These data suggested that either E-selectin plays only a minor role in the extravasation of neutrophils or that an additional molecule can compensate for the loss of E-selectin function. In order to see whether the extravasation in E-selectin-deficient mice relies on the remaining endothelial selectin, P-selectin, we examined neutrophil accumulation in thioglycollate-induced peritonitis after blockade of P-selectin function using a newly characterized anti-murine P-selectin monoclonal antibody, 5H1 (manuscript in preparation). Wild-type and E-selectin −/− mice were treated with thioglycollate or with thioglycollate and 5H1 (intravenous administration). The numbers of neutrophils were examined at either 2 or 6 h after injection. As shown in Fig. 2, neutrophil accumulation at 2 h was dependent on P-selectin in both wild-type and E-selectin-deficient mice. 5H1 blocked influx of neutrophils to an equivalent level in both groups. The anti-P-selectin antibody had no effect, however, at 6 h in wild-type mice, supporting the notion that, in this model, an additional molecule substitutes at later times for the loss of P-selectin function. In contrast to wild-type mice, neutrophil accummulation was efficiently inhibited by the anti-P-selectin antibody in the E-selectin-deficient mice. This proves that the majority of neutrophil migration in the E-selectin −/− animals is mediated by P-selectin and that E-selectin is responsible for mediating neutrophil influx in the absence of P-selectin function in the wild-type mice at late times after injection with thioglycollate.

Taken together, the studies with the wild-type and E-selectin −/− mice suggest that neutrophil migration can be mediated by either of the endothelial selectins and that E- and P-selectin are functionally redundant. The absence of a requirement for E-selectin function at early times in this model of inflammation is likely due to the fact that there is little expression of E-selectin at these times. Although expression of both murine P-selectin and E-selectin appears to be induced by cytokines (Fig. 1 and Weller et al 1992), P-selectin protein is pre-formed and stored in Weibel–Palade bodies, and is rapidly released and transported to the cell surface upon activation of endothelial cells. Thus the influx of leukocytes early on in inflammatory reactions is likely dependent on P-selectin, because that is the predominant or only selectin on endothelial cells at those times. The results presented in this paper strongly support an essential role for endothelial selectins in leukocyte extravasation. Previous studies in which the activity of only single selectins was blocked often resulted in only a partial inhibition of neutrophil immigration (Mayadas et al 1993, Mulligan et al 1991, Watson et al 1991). The data presented here suggest that the partial inhibition seen in previous experiments was due to compensation by other selectins. In this regard, it is important to note that although 5H1 blocked approximately 90% of neutrophil influx in the context of E-selectin-deficient mice, some neutrophils still migrated into the peritoneum. This result may have been a consequence of incomplete blockade by 5H1. Alternatively, the remaining neutrophil influx may have been a result of the interaction between L-selectin present on neutrophils and a co-receptor on endothelial cells. The contribution of L-selectin in P- and E-selectin-independent neutrophil influx can now be readily tested using anti-L-selectin antibodies with the reagents presented in this paper. Finally, this work demonstrates that selectin expression on endothelial cells plays a major role at all times in a model of acute inflammation. The notion that E- and P-selectin may be functionally redundant suggests that therapeutically useful antagonists may need to inhibit both E-selectin and P-selectin.

Characterization of VCAM-1-deficient mice

VCAM-1 was originally isolated as a cytokine-inducible adhesion molecule expressed on human umbilical vein endothelial cells which could mediate their binding to a number of leukocytes, including B cells, T cells, mast cells and monocytes, through an interaction with the integrin $\alpha_4\beta_1$ (also known as VLA-4) (Osborn et al 1989, Elices et al 1990). VCAM-1 expression has been associated with a number of disease states in humans (Rice et al 1991) and antibody to VCAM-1 has been shown to prevent the rejection of cardiac allografts in mice (Pelletier et al 1992). In addition to its role in mediating inflammatory disease, it has been suggested that VCAM-1 contributes to a number of normal immunological and developmental processes. VCAM-1 has previously been shown to be expressed constitutively in bone marrow stromal cells, dendritic cells of lymphoid organs and developing skeletal muscle (Miyake et al 1991, Rosen et al 1992).

In order to develop a model to study the potential roles of VCAM-1 in inflammatory disease and/or in development, we introduced a null mutation into the VCAM-1 gene in ES cells. The genomic structure of the murine VCAM-1 locus was described by Terry et al (1993) and is shown in Fig. 3 along with the structure of the VCAM-1 targeting vector. The details of the construction of the targeting vector will appear elsewhere. The VCAM-1 gene contains separate exons encoding each Ig domain. An alternatively spliced VCAM-1 RNA, produced after cytokine induction in endothelial cells, encodes only the first three Ig domains and an alternative exon 5 (Δ5) encoding the addition site for a glycosylphosphatidylinositol anchor. The targeting vector introduced two mutations in the VCAM-1 gene. First the *neo*^{r} gene was inserted within the exon encoding domain I, resulting in a large insertion/frameshift mutation. Second, a deletion was introduced spanning sequences from exons encoding domain I to domain IV. Thus, the mutation removed all Ig domains capable of interaction with $\alpha_4\beta_1$. Chimeric animals were produced with two ES cell clones containing the targeted mutation and were bred to see if the VCAM-1-deficient animals were viable. As shown in Table 1, after extensive breeding only one live animal homozygous for the VCAM-1 mutation was detected after screening of more than 700 offspring from inter-crosses between heterozygous animals. The one homozygous animal died at 6 weeks of age and its remains were not available for analysis. Thus, it can be concluded that the

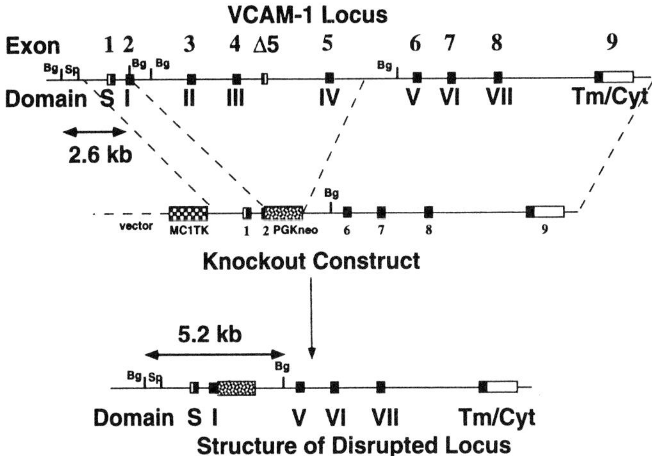

FIG. 3. Disruption of the murine vascular cell adhesion molecule (VCAM)-1 gene. The structure of the murine VCAM-1 gene is shown along with the VCAM-1 knockout construct and structure of the targeted locus. Also shown are the location of restriction enzyme *Bgl*II sites used to identify mice carrying the mutated allele. A detailed description of the constructs and targeting in embryonic stem cells will be presented elsewhere.

VCAM-1 gene is essential for normal development. In order to determine when the VCAM-1 gene was required, we isolated embryos from inter-crosses at various times, and we determined their genotypes by Southern blotting of yolk sac DNA (data not shown). The majority of VCAM-1 −/− embryos died and were resorbed by 12.5 d gestation. A small number of normal-sized apparently live −/− embryos were found at 12.5 d (2 out of 23 homozygotes) and were used for analysis (see below). Analysis of earlier embryos suggested that the majority of −/− animals die between 10.5 and 12.5 d. At 10.5 d most −/− mice were relatively normal, although half were already reduced in size. By 11.5 d 32% of the −/− embryos have died. The dead embryos were highly necrotic, very small and pale. However, the majority of −/− embryos were of normal size and appearance at 11.5 d. These observations suggested that the VCAM-1 embryos exhibit two distinct phenotypes. One set of embryos died before 11.5 d while the other set remained viable for another day.

In order to predict which developmental processes were affected by the VCAM-1 mutation, we examined the sites of VCAM-1 expression in 8.5 d wild-type embryos by wholemount immunohistochemistry. These experiments revealed two major sites of VCAM-1 expression. As shown in Fig. 4, a high level of VCAM-1 protein was detected within the distal end of the allantois, an extraembryonic tissue that establishes contact between the embryo and the placenta by fusion to the chorion. It is essential both for respiration and waste elimination by the rapidly growing embryo. High levels of VCAM-1 expression

TABLE 1 Genotypes of progeny from VCAM-1$^-$/VCAM-1$^+$ intercrosses

	+/+	+/−	−/−
Live births	208 (36%)	363 (63%)	1[a] (0.2%)
10.5 dpc	15 (25%)	32 (53%)	13 (22%) (6/13 small)
11.5 dpc	21 (28%)	35 (47%)	13 live[b] (17%) 6 dead (8%)
12.5 dpc	21 (25%)	39 (47%)	2 live (2%) 21 dead (25%)

[a]The one homozygous −/− adult died at 6 weeks of age and the remains were not available for analysis. Thus its identity as a −/− animal could not be verified.
[b]Embryos were judged as live if they were normal in size and colour and if the heart appeared to be beating.

were also observed in the developing heart. This expression persisted until at least 12.5 d, although it gradually became more restricted (S. Baldwin, unpublished data). Other experiments have demonstrated that this expression was restricted to the myocardium of the developing heart and was particularly high in the ventricular septum (data not shown). The rapid death of the embryos (Table 1) suggested that the VCAM-1 deficiency may have affected the development of a tissue essential for the growth of the entire embryo. Defects in either the development of the heart or in the allantois could have severe effects on embryo development and survival. Thus development of these tissues was examined in detail.

In order to determine if development of the allantois was affected by the VCAM-1 mutation, we isolated and examined 9.5 d embryos. Dissections were carried out such that the embryos were recovered in intact extraembryonic membranes. After photographing the embryos, we dissected them and used the membranes to prepare DNA for genotyping. Although all of the 9.5 d embryos were of normal size and appearance, many of the VCAM-1 −/− embryos had defective allantoic structures, as shown in Fig. 4. In general, the defect appears to be a failure of the allantois to fuse with the chorion. Approximately 50% of the −/− embryos had large fluid-filled allantoic stalks or a smaller-than-normal necrotic allantois. Some −/− embryos contained an allantois that was loosely attached to the chorion. This attachment was often not at the normal, most distal, end of the allantois, but was on the side of the allantois. The defective allantoic structures are sufficient to explain the rapid deaths of a significant number of VCAM-1 −/− embryos and explain why approximately half of the 10.5 d embryos are smaller than wild-type, as shown in Table 1. The small 10.5 d −/− embryos most likely failed to establish a connection to the placenta. However, a large proportion (70%) of 11.5 d and a very small number of 12.5 d embryos were relatively normal in appearance. In fact, the majority

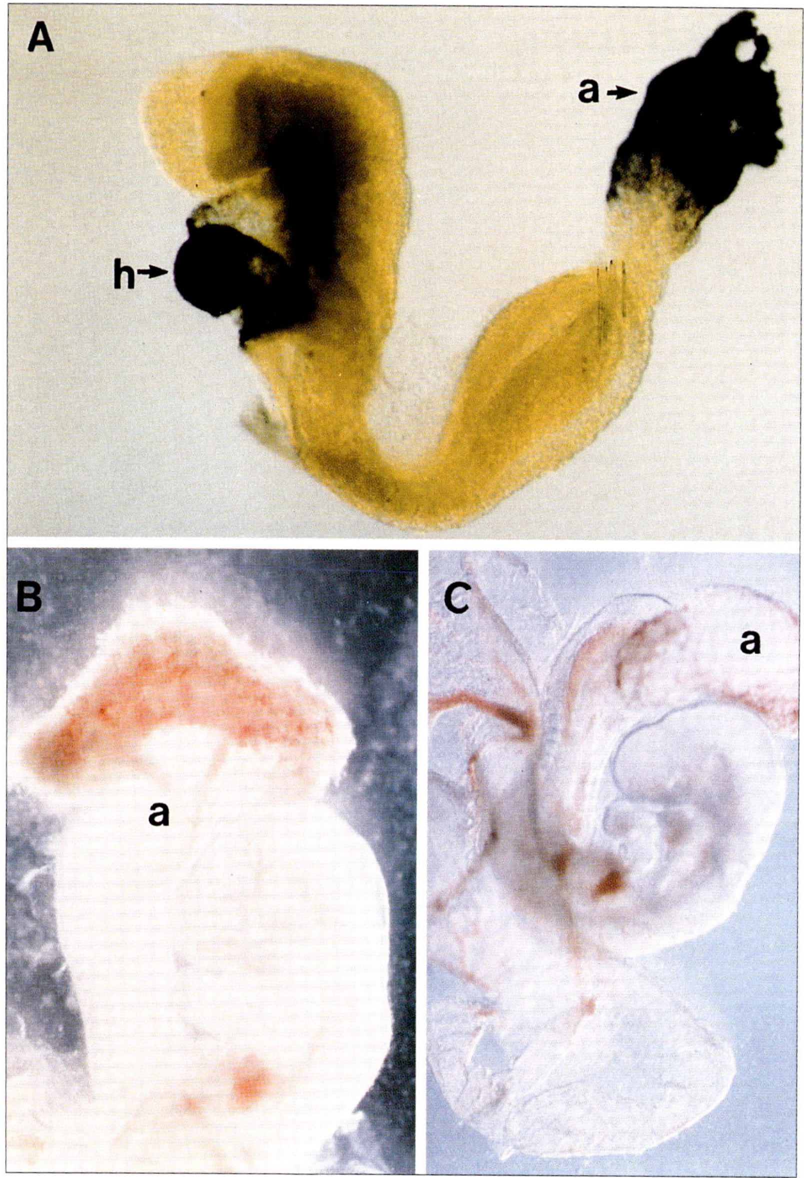

FIG. 4. Expression of vascular cell adhesion molecule (VCAM)-1 in early mouse embryos and the effect of VCAM-1 gene disruption on the development of the allantois. (A) An 8.5 d embryo subjected to wholemount immunohistochemical analysis using an anti-murine VCAM-1 antibody is shown. Specific expression was detected in the heart (h) and in the distal end of the allantois (a). No staining in these tissues was observed when experiments were carried out without the primary antibody (data not shown). (B) Photomicrograph of a wild-type 10.5 d mouse embryo. The location of the allantois is marked with an (a) on the left. The allantois appears normal and is clearly fused with the placenta (red). (C) Photomicrograph of a VCAM-1 -/- litter mate of the embryo shown in (B). As can be readily seen, the embryo contains a swollen allantois (a) which has failed to fuse to the placenta.

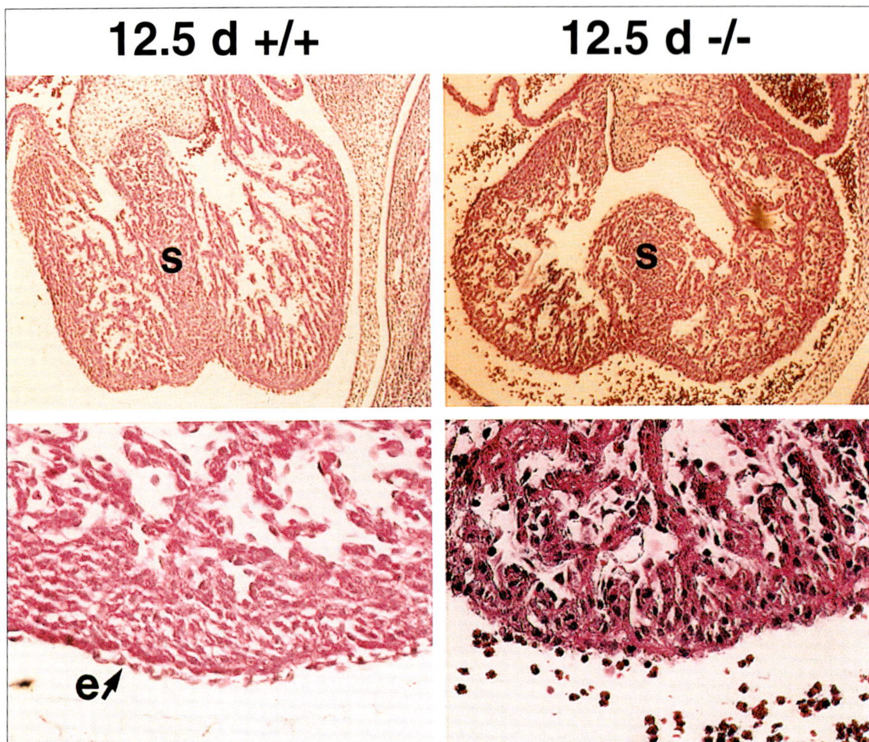

FIG. 5. Hearts of the VCAM-1 -/- embryos contain multiple defects. Sections of the cardiac ventricles from 12.5 d +/+ and -/- embryos are shown at two different magnifications (top panels x 100, bottom x 400). The location of the ventricular septa (s) and the epicardium (e) of the +/+ animal are shown. The VCAM-1 -/- embryo had no discernible layer of epicardial cells. Note the reduced thickness of the myocardium of the -/- embryo particularly as seen in the higher power magnification (lower right). Also note the large amount of blood within the pericardial sac of the -/- heart.

of surviving 11.5 d −/− embryos appeared to have functional umbilical arteries and veins as well as normal yolk sac circulation (data not shown) and thus seem to have established an intact extraembryonic circulation. It is possible, however, that although the umbilicus in these embryos has developed, it is not properly fused with the chorion and still produces a defective circulatory system. The extraembryonic structures of the VCAM-1 −/− embryos are currently being examined in more detail.

We also examined relatively normal 11.5 and 12.5 d VCAM-1-deficient embryos histologically to see if the VCAM-1 mutation affected development of the heart. Analysis of the most healthy looking embryos revealed four distinct defects in the hearts of VCAM-1 −/− embryos (Fig. 5). At 12.5 d, these hearts had a thinner ventricular myocardium; this was particularly pronounced in the compact zone of the ventricle. The ventricular myocardium also appeared less dense and more disorganized. In addition, the ventricular septa of both 11.5 (not shown) and 12.5 d embryos were significantly reduced in size compared with wild-type litter mates. The reduction in size and thickness of myocardium was observed in most −/− embryos but varied greatly in degree. All of the hearts from 11.5 and 12.5 d VCAM-1 −/− embryos contained large pericardial effusions of blood and very little or no epicardium.

The most profound defect in the VCAM-1-deficient embryonic hearts was a failure of the epicardium to form: no epicardial layer could be seen, while the wild-type litter mates clearly had a layer of flattened cells surrounding the myocardium. Some possible epicardial cells were observed at very reduced numbers in some −/− embryos, particularly on the atrial myocardium (data not shown), but in all cases represented a largely incomplete covering of the myocardium. Recently, we and others (data not shown and Sheppard et al 1994) have demonstrated that $\alpha_4\beta_1$ is expressed within the epicardium. Additional experiments (manuscript in preparation) failed to detect any $\alpha_4\beta_1$-positive cells lining the myocardium of the −/− mice, supporting the histological data suggesting that the epicardium is missing.

There are several explanations for these observations. First, the defective extraembryonic circulation may indirectly and preferentially affect the growth of these heart structures. This seems unlikely, since rapidly growing structures in the heart, particularly in the atrium, were not affected in the VCAM-1 −/− embryos. Alternatively, VCAM-1 may be important for the growth of the ventricular myocardium. In this regard, it should be noted that although VCAM-1 is expressed throughout the myocardium, the highest levels were observed specifically in the ventricular septum and in the compact zone of the myocardium (data not shown). A third explanation is that the lack of an epicardium results in the more subtle changes occurring in the ventricle. This seems most likely for several reasons. First, the epicardium was missing in all 11.5 and 12.5 d −/− embryos examined, and although the ventricular myocardium and septum were reduced in size, they were not absent and the extent of this size reduction was variable. Thus the growth of the myocardium and septum may have been indirectly retarded. Secondly, as the epicardium is thought to provide a protective role for the heart, the loss of this cell layer may explain the large pericardial effusion seen in all $-/-$ hearts. The bleeding into the pericardial sac may be the result of a decrease in the structural integrity of the heart and could lead to reduced cardiac output. In addition, the epicardium may play a role in the growth or function of the myocardium by secretion of growth or regulatory factors. Finally, a role for VCAM-1 $\alpha_4\beta_1$ interaction in the development of the epicardium is mechanistically attractive. The epicardium is thought to form through invasion of the heart by mesothelial cells which attach to the outer layer of the myocardium and differentiate into the epicardium. The data presented here suggest a model where VCAM-1 would act as a homing receptor for migrating $\alpha_4\beta_1$-positive epicardial precursors. The lack of this homing receptor would be predicted to prevent or greatly diminish the colonization of the heart by the mesothelial cells. This model is directly analogous to the role of VCAM-1 and $\alpha_4\beta_1$ in mediating leukocyte trafficking in the adult circulation. On the other hand, a direct role for VCAM-1 in the organization or growth of the myocardium would require the existence of unknown VCAM-1 ligands or additional functions other than cell adhesion.

Conclusions

This work describes the creation and characterization of lines of mice deficient in expression of specific adhesion molecules. These mice differ dramatically in their phenotypes. E-selectin-deficient mice were viable while VCAM-1-deficient mice died between 10.5 and 12.5 d of gestation. These data suggest that E-selectin has evolved specifically to facilitate trafficking of leukocytes. This conclusion is consistent with the proposed role of the selectins as molecules specialized for mediating the interaction of leukocytes with the vessel wall under conditions of shear forces seen in normal blood flow. It is also possible, however, that just as the role of E-selectin in mediating neutrophil migration overlaps with that of P-selectin, any role it has in development may also be complemented by other selectins, although this seems unlikely since there is no clear evidence for E-selectin expression in the absence of inflammatory stimuli. Analysis of the E-selectin mutants clearly demonstrates a major role for the selectins in neutrophil migration. It will be important for us to carry out similar double-knockout experiments (or triple-selectin knockout experiments) in a number of disease models in order to determine if the requirement for multiple selectins is a general property of neutrophil/leukocyte emigration.

In contrast to E-selectin-deficient mice, VCAM-1-deficient mice were not viable. This result is not surprising in that, unlike the selectins, VCAM-1 is expressed in a number of developing tissues in the absence of inflammatory stimuli. Initial examination of the VCAM-1 $-/-$ embryos demonstrated that at least two tissues were affected by the mutation. The earliest observed embryonic phenotype was a defective allantois. The embryos that survived the longest also contained specific cardiac defects, most prominent of which was the loss of the epicardium. The complex phenotypes of VCAM-1 mutants as well as the intricate patterns of expression suggest that the VCAM-1 gene has evolved to carry out specific developmental functions and may have later been adapted for use in the immune system. Unfortunately, the VCAM-1-deficient mice described in this work do not provide a useful model for studying the role of VCAM-1 in mediating inflammatory disease. The data presented in this paper should, however, help us to design experiments to produce adult VCAM-1-deficient animals useful in immunological studies by either rescuing the developmental lethality of VCAM-1 deficiency using VCAM-1 transgenes or by selectively disrupting VCAM-1 expression in endothelial cells.

Acknowledgements

We would like to thank Doug Larigan, Joe Levine, Lucy Foppiani, Rongshen Hsiao and John Duker for oligonucleotides and sequencing. We are also grateful to Dr Greg Pirozzi and Gwen Wong for helpful discussions.

References

Abbondanzo SJ, Gadi I, Stewart CL 1993 Derivation of embryonic stem cell lines. Methods Enzymol 225:803–823

Becker-André M, Van Huijsduijnen RH, Losberger C, Whelan J, Delamarter JF 1992 Murine endothelial leukocyte-adhesion molecule 1 is a close structural and functional homologue of the human protein. Eur J Biochem 206:401–411

Bevilacqua MP 1993 Endothelial–leukocyte adhesion molecules. Annu Rev Immunol 11:767–804

Elices MJ, Osborn L, Takada Y et al 1990 VCAM-1 on activated endothelium interacts with the leukocyte integrin VLA-4 at a site distinct from the VLA-4/fibronectin binding site. Cell 60:577–584

Mayadas TN, Johnson RC, Rayburn H, Hynes RO, Wagner DD 1993 Leukocyte rolling and extravasation are severely compromised in P selectin-deficient mice. Cell 74:541–554

Miyake K, Medina K, Ishihara K, Kimoto M, Auerback R, Kincade PW 1991 A VCAM-like adhesion molecule on murine bone marrow stromal cells mediates binding of lymphocyte precursors in culture. J Cell Biol 114:557–565

Mulligan MS, Varani J, Dame MK et al 1991 Role of endothelial–leukocyte adhesion molecule 1 (ELAM-1) in neutrophil-mediated lung injury in rats. J Clin Invest 88: 1396–1406

Osborn LC, Hession R, Tizard C et al 1989 Direct expression cloning of vascular cell adhesion molecule 1, a cytokine-induced endothelial protein that binds to lymphocytes. Cell 59:1203–1211

Pelletier RP, Ohye RG, Vanbuskirk A et al 1992 Importance of endothelial VCAM-1 for inflammatory leukocytic infiltration in vivo. J Immunol 149:2473–2481

Rice GE, Munro JM, Corless C, Bevilacqua MP 1991 Vascular and nonvascular expression of INCAM-110. Am J Pathol 138:385–393

Rosen GD, Sanes JR, LaChance R, Cunningham JM, Roman J, Dean DC 1992 Roles for the integrin VLA-4 and its counter receptor VCAM-1 in myogenesis. Cell 69:1107–1119
Sheppard AM, Onken MD, Rosen GD, Noakes PJ, Dean DC 1994 Expanding roles for α4 integrin and its ligands in development. Cell Adhes Comm, in press
Springer TA 1994 Traffic signals for lymphocyte recirculation and leukocyte emigration: the multistep paradigm. Cell 76:301–314
Stewart CL, Kaspar P, Brunet LJ 1992 Blastocyst implantation is dependent on maternal expression of leukemia inhibitory factor. Nature 358:76–80
Terry RW, Kwee L, Levine JF, Labow MA 1993 Cytokine induction of an alternatively spliced murine vascular cell adhesion molecule (VCAM) mRNA encoding a glycophosphatidylinositol-anchored VCAM protein. Proc Natl Acad Sci USA 90:5919–5923
Thomas KR, Capecchi MR 1987 Site-directed mutagenesis by gene targeting in mouse embryo-derived stem cells. Cell 51:503–512
Watson SR, Fennie C, Lasky LA 1991 Neutrophil influx into an inflammatory site inhibited by a soluble homing receptor-IgG chimera. Nature 349:164–167
Weller A, Isenmann S, Vestwever D 1992 Cloning of the mouse endothelial selectins: expression of both E- and P-selectin is induced by tumor necrosis factor. J Biol Chem 267:15176–15183

DISCUSSION

Hynes: Our integrin $\alpha_4\beta_1$ (the VCAM-1 receptor) knockout mice have a similar phenotype to the VCAM-1 knockouts you have described, with some variations (J.T. Yang, H. Rayburn, R.O. Hynes, unpublished results). The placental defect and fusion defect are exactly the same and, as you mentioned, $\alpha_4\beta_1$ is expressed on the chorion. The $\alpha_4\beta_1$ knockout also lacks an epicardium, which forms from the septum transversum and migrates over the surface of the developing heart. When Joy Yang, in our laboratory, first saw the absence of epicardium we thought, as you have done, that this was because of a failure in migration or homing. This is not the case. If you look at 10.5 d, the epicardium is present even in the mutants. By 11.5–12.5 d it has gone, so it disappears rather than fails to get there. One possibility is that it gets sloughed off when the heart beats.

Joy Yang showed that $\alpha_4\beta_1$ is present particularly in the epicardium and the endocardial cushions. If you stain for fibronectin, you see exactly the same predominance. $\alpha_4\beta_1$ is a fibronectin receptor as well as a VCAM-1 receptor. So, by looking at expression you can't tell whether the defect is due to the failure of $\alpha_4\beta_1$ integrins to bind to VCAM-1, or to fibronectin, or both. The fibronectin mutant dies before this stage, so there is no genetic information about it. Consequently, we're still left with a puzzle as to the nature of the cardiac defect.

Labow: The similarity of the two mutants would suggest that, in development, it is the $\alpha_4\beta_1$–VCAM interaction that is most important.

Hynes: I grant you that. I just raised the possibility that fibronectin could also be involved, because it is also present. There's a contrary sort of puzzle in the endocardial cushions where there is no VCAM, but $\alpha_4\beta_1$ is present, and the cushions develop fine in the α_4 mutants. There, as one often sees, expression gives you a misleading answer. The fact that $\alpha_4\beta_1$ is present doesn't mean it's essential.

Birchmeier: I was astonished that in the VCAM-1-deficient mice you were disappointed to see a developmental defect rather than a defect in inflammation: I always find the developmental defects particularly interesting. Could you rescue your mouse by expressing VCAM-1 in the heart?

Labow: We're trying to do that. This is the only way we're going to be able to prove that the heart defect is a primary defect. If there are subtle defects in the extraembryonic circulation the mouse might be subsequently compromised so that heart development is affected. We're trying to do this rescue in two ways: one is by expression of VCAM-1 transients and the other is by creating site-specific recombination systems. We've already engineered the constructs. What we really want to get is an adult animal that lacks the ability to make VCAM-1 in inflammatory scenarios. The rescue is very difficult because of the allantois defects, and I know of no promoters that I can use to drive expression to the distal end of the allantois.

Humphries: One possible function for VCAM-1 is to modulate haemopoietic stem cell differentiation. In your homozygous VCAM-1 knockout embryos, can you rescue blood cells to look at the different levels of subtypes and therefore see if a lack in VCAM-1 affects this problem?

Labow: I think we could, but we haven't. We probably won't have the time to do it.

Hynes: We have made chimeras that are a mixture of $\alpha_4\beta_1$-positive and $\alpha_4\beta_1$-negative cells by making a double knockout ES cell, then making chimeras with that. These animals are viable: we can look at the function of the $\alpha_4\beta_1$-negative cells in their circulation. It's easier there because we're looking at a circulating cell population. In your case, it would be harder—you'd have a patchy endothelium—but it's still feasible.

Labow: Unfortunately, we tried to produce VCAM-1 $-/-$ ES cells and we were unlucky. We had the wrong *neo* resistance cartridge in the knockout construct, which prevented us from using gene conversion to disrupt both VCAM alleles.

Wagner: The allantoic defect in the VCAM-1 knockout mice seems to be 'leaky' in the sense that 30% of the animals survive 10.5 d. Richard Hynes, how is it in your $\alpha_4\beta_1$ knockouts?

Hynes: Fewer than half of them die at the placental defect stage, and the other half, in our experience, go on further than Mark Labow's. Several of ours get through to Day 14 or so. I think our heart defects are somewhat different—we don't see a defect in the septum, for instance.

The fibronectin knockouts also have two different phenotypes: this turns out to be due to genetic background, because they never form a heart in the 129 background and they always form a heart in the C57BL background. But in these $\alpha_4\beta_1$ knockouts (and presumably the same is true for the VCAM-1 knockouts) there's a mixture (of phenotypes), and we don't yet know for sure whether it's due to the genetic background or not. Joy Yang is now looking at pure 129 litters, which are easier to get than pure C57 litters, and it still looks mixed. The preliminary data suggest that it's not due to genetic background. It could be stochastic, it could be that there are several adhesion molecules involved in the allantois–chorion fusion: it's just a matter of how far down that path you've got before the animal turns. Mouse embryos do a sort of somersault around this time when the fusion is supposed to be happening and if they achieve the somersault before the fusion, then there's no way the allantois is ever going to find the chorion.

You suggested that E-selectin and P-selectin might be redundant as a pair. I think it is helpful to distinguish between *redundancy* and *compensation*. I would define redundancy as the situation where both molecules are normally present; then, if you knock one out, the other one suffices, whereas compensation, which would produce the same end result, would be the situation where one of them is there, you knock it out and the other one then replaces it by up-regulation. Do you have any evidence as to which of these situations is occurring in your E-selectin knockouts?

Labow: E- and P-selectin are not exactly redundant because, clearly, P-selectin is important earlier, as its regulation is somewhat different. We think they're functionally redundant. We looked very carefully at P-selectin expression in the E-selectin knockouts: sometimes it's twofold higher and sometimes it's twofold lower compared with the wild-type animals. One of the problems is that our work is predominantly on F2 hybrids between C57 and 129 mice. Only now are we doing the experiments on C57BL/6 backgrounds (made for doing metastasis studies) and 129 backgrounds. The genetic variability may explain why there's a lot of flux in our results and the new experiments should give us cleaner answers on the changes in regulation. Consistent with this, expression of P- and E-selectin in different strains of mice in terms of cytokine responsiveness can be very different. For example, FVB mice are very good cytokine responders and BL/6 mice are not so good. Slight genetic shuffling might give you subtle differences that make the numbers hard to interpret. If the selectins are functionally redundant, it doesn't mean that they're interacting with the same ligands, just that for the inflammatory response either one seems to be sufficient.

Ruggeri: Human genetic studies might be fairly telling with regard to P- and E-selectin, because you would expect that if one can function instead of the other, there should be an accumulation of mutations in these two genes in humans. With platelet proteins that are obviously not necessary for life, such as GPIb and GPIIb/IIIa, typically you have a fairly large number of mutations.

Hynes: There's an assumption there (which is probably not valid) that just because these mice aren't dead or really sick, there's nothing wrong with them. If the mutation were in a person, you would see them presenting with infections or other problems.

Ruggeri: I'm just asking about what is known in human population genetics about P- and E-selectin.

Wagner: I don't think very much is known at all. The problem may be that these are very recently uncovered markers.

Hynes: What are the major diseases mapped to that arm of that chromosome?

Wagner: I don't know.

Etzioni: The LAD II patients are not very severely infected (Etzioni et al 1995, this volume). If they didn't have such a high leukocyte count, I don't think we would have discovered them, aside from other phenotypic characteristics they exhibit that don't have to do with cell adhesion. Maybe there *are* people who are selectin deficient and who we fail to identify because they don't have a severe phenotype.

Ruggeri: It might be helpful for us to figure out what these molecules do in humans, because studying just one species could be misleading. For example, aspirin is one of the most effective anti-platelet drugs in humans, but it does absolutely nothing in rodents because their platelets can use other pathways for activation. Human studies would certainly be very useful to complement the mouse studies.

Garrod: Perhaps before you do that you ought to give the mice more of an environmental challenge.

Ruggeri: Yes, but there could be significant species differences.

Labow: You have to take the lessons that we've learned with mice with several grains of salt, because we are shooting these animals with inflammatory cannonballs, whereas, in humans, the usual diseases are initiated by much more subtle inflammatory cues such as insect bites and poison ivy.

Wagner: What about sepsis?

Labow: Sepsis would be much more akin to what we're looking at with mediators like thioglycollate and LPS.

Barker: Eugene Butcher's group has shown that E-selectin expression has organ specificity: there's more of it in the skin, for example, than in the joints (Picker et al 1993). Dorian Haskard has evidence that it is strongly expressed in pig skin (Keelan et al 1994, Jamar et al 1994). What is the situation with mice? If you're looking at the mouse as a model of human disease with respect to E-selectin, does E-selectin have a similar pattern of expression in mice to that in humans, or should you be looking at another animal model?

Labow: There are reasons the mouse is a relevant model. It has superior genetics—we can't do these experiments in any other animal. But, I agree, in looking at a very specific disease you want to go back and make sure the regulation is similar. There is some similarity: E-selectin is expressed in the skin.

We can see it in the small venules. There are very high levels of E-selectin in the liver. There's a high level of expression in kidney. That's all the information we really have. The problem is that we can't do the same experiments in humans as in the mouse.

Wagner: Did you look at the constitutive expression of E-selectin? Is it ever constitutively expressed?

Labow: No, we didn't, but we will go back and do it. One of the points you're getting at is whether the selectins are required in development. If our interpretation that the functions of P- and E-selectin are overlapping is reasonable, then the observation that individual mice are viable or fertile may not be relevant, and the same situation might be occurring in development. I don't know of any studies that have looked for selectins in development. The only information we have is that E-selectin is expressed in the uterus during implantation; we don't know in which cell types. Expression of E-selectin may be a consequence of cytokine production during implantation. It's very similar to the inflammatory process—VCAM-1 and E-selectin are both expressed and you see high levels that seem to spike right around implantation, at about 3.5 d (these are experiments that we're doing with Colin Stewart).

Etzioni: We recently had a LAD II female fetus aborted at 20 weeks' gestation (Frydman et al 1994). She was completely normal microscopically and macroscopically. From this case it seems that E-selectin doesn't play a crucial role. Do you have a high neutrophil count in your E-selectin knockout mice, just as we have in our LAD II patients?

Labow: Yes, they do seem to have a mild neutrophilia. I don't think it's nearly as significant as that seen in the P-selectin-deficient animals, which would concur with Denisa Wagner's data that spontaneous rolling is a P-selectin-mediated process (Wagner 1995, this volume). Usually, we see an increase of about 70%. If you take the E-selectin knockouts and induce inflammation by thioglycollate and give the mice anti-P-selectin antibody, not only do you prevent neutrophil trafficking, you also see an enormous increase in the numbers of peripheral neutrophils. It's much larger than the increase we see in just the E-selectin background and it's probably much larger than in the P-selectin background. Denisa Wagner, do you see similar things when you treat your P-selectin knockout mice with thioglycollate? At times when you see a decrease in neutrophil recruitment, do you see a comparable increase in the periphery?

Wagner: Yes, we see an increase in the number of peripheral neutrophils in general.

Labow: Does it correlate with just the neutrophils that are missing in the peritoneum? Our numbers seem larger than this can account for.

Wagner: It is possible that your numbers are larger because they also include the marginating pool of neutrophils.

Labow: It's very hard to say what this means, but it suggests to us (and this is the kind of thing you put at the very bottom of a paper, if you get away with it) that the selectins might be involved in sequestering these marginating pools, because when we block all the selectins we do see a large increase in the number of neutrophils.

Pober: If you want to uncover functional defects in E-selectin, it would be advantageous to start with the same models that Peter Ward and his colleagues have used to implicate E-selectin by means of antibody blocking experiments (reviewed in Albelda et al 1994). One prediction from those experiments is that non-specific peritonitis is not likely to be inhibited by antibody to E-selectin. The clear cases in which E-selectin effects were observed are immune complex-induced inflammation, particularly passive reverse Arthus reactions in the lung or skin. Have you looked at those models?

Labow: We have not studied any immune complex disease models, although we are trying to get various forms of these models going. I think that E-selectin does have a role in thioglycollate action, but it's only uncovered if you remove P-selectin. Clearly, Denisa Wagner's data agree with ours: if you lose P-selectin, you still have relatively normal rates of neutrophil recruitment at late times. We have looked at some models of lung inflammation, in which we give intranasal applications of LPS. There we see results similar to those in thioglycollate-induced peritonitis, although blocking both selectins is not nearly as efficacious: you get a good 50–60% reduction in neutrophil accumulation, but you can't bring it down to the baseline as you can with peritonitis and delayed-type hypersensitivity. Perhaps it's not selectin dependent but is instead a consequence of changes in flow rate of blood due to physical damage to the lung.

Stanley: I have a question about margination, because I don't think it's been adequately addressed. Margination of neutrophils (at least in humans) is a very dramatic effect. It's said that at least 50% of the neutrophils are marginated at any one time. Do we know about margination? Do we know if selectins are involved? In Denisa Wagner's P-selectin knockout (Wagner 1995, this volume) there was very little effect on margination. Margination is affected dramatically by corticosteroids; if you give a patient corticosteroids, their neutrophil count will go up in a few hours.

Wagner: Basically, we know nothing about margination. There may be several pools of neutrophils that are hidden when you draw blood and you do peripheral neutrophil counts, and they may be in totally different places. A lot of investigators think that there is a lung pool of neutrophils in capillaries. When you give adrenaline, these small vessels may dilate and let the neutrophils out through physical means alone. Another possible marginating pool comprises the rolling leukocytes. Whether this pool exists in an unstimulated animal and whether adrenaline prevents rolling are not known. All I can say is that when we give adrenaline to P-selectin-deficient animals, there is a pool of neutrophils that returns into the bloodstream and, in absolute value, this pool of neutrophils is identical to that observed in the wild-type animals. Therefore, at least this pool is not bound somewhere through P-selectin.

Hynes: One of the problems with discussing 'margination' is the use of that word; it contains the assumption that we understand where that pool is resting, that it's rolling on the wall (the 'margin'). I don't know of any evidence (it may be my ignorance) that that is what it is. The 'margination' is usually detected by something like adrenaline or corticosteroid, which could be releasing the pool from some other source. It might be much easier to think about this issue if we stop using the loaded terminology.

Shaltiel: You blocked the functions of the various selectins by using E-selectin knockout mice and antibodies against the other selectins. Could you achieve such blocking with soluble forms of both selectins? This would demonstrate competitiveness and help elucidate the contribution of the transmembrane piece.

Labow: I think we could, and certainly this has been done to some extent. Watson et al (1991) at Genetech have made IgG chimeras and a number of people have shown that L-selectin–IgG chimeras are actually inhibitory. We found that the problem with the soluble E-selectin molecules is that they're somewhat low-affinity receptors and we need really high concentrations to tether leukocytes. The other advantage in using IgG chimeras and antibodies is that they seem to last much longer in the circulation.

References

Albelda SM, Smith CW, Ward PA 1994 Adhesion molecules and inflammatory injury. FASEB (Fed Am Soc Exp Biol) J 8:504–512

Etzioni A, Phillips LM, Paulson JC, Harlan JM 1995 Leukocyte adhesion deficiency (LAD) II. In: Cell adhesion and human disease. Wiley, Chichester (Ciba Found Symp 189) p 51–62

Frydman M, Verdimon D, Sholer E, Orlin GB 1994 Prenatal diagnosis of LAD II syndrome. Prenatal Diagn, in press

Jamar F, Chapman PT, Harrison AA, Binns RM, Haskard DO, Peters AM 1994 Imaging endothelial cell activation in inflammatory arthritis using an In-111 labeled anti-E-selectin monoclonal F(ab')2. Radiology, in press

Keelan ETM, Licence ST, Peters AM, Binns R, Haskard DO 1994 Characterization of E-selectin expression in vivo using a radiolabelled monoclonal antibody. Am J Physiol 266:H279–H290

Picker LJ, Michie SA, Rott LS, Butcher EC 1993 A unique phenotype of skin-associated lymphocytes in man: preferential expression of HECA-452 epitope by benign and malignant T-cells at cutaneous sites. Am J Pathol 136:1053–1061

Wagner DD 1995 P-selectin knockout: a mouse model for various human diseases. In: Cell adhesion and human disease. Wiley, Chichester (Ciba Found Symp 189) p 2–16

Watson SR, Fennie C, Lasky LA 1991 Neutrophil influx into an inflammatory site inhibited by a soluble homing receptor-IgG chimera. Nature 349:164–167

Von Willebrand's disease and the mechanisms of platelet function

Zaverio M. Ruggeri

Roon Research Center for Arteriosclerosis and Thrombosis, Division of Experimental Thrombosis and Hemostasis, Departments of Molecular and Experimental Medicine and of Vascular Biology, The Scripps Research Institute, 10666 North Torrey Pines Road, La Jolla, CA 92037, USA

Abstract. Von Willebrand's disease, the most common congenital bleeding disorder in humans, is the consequence of quantitative and/or qualitative defects of von Willebrand factor, a protein necessary for platelet adhesion and thrombus formation at sites of vascular injury. The definition of the molecular basis of von Willebrand's disease has helped clarify the structure of von Willebrand factor as well as its essential role in platelet function, particularly under haemodynamic conditions of high shear stress. Platelets respond rapidly to alterations of endothelial cells by attaching firmly to the site of lesion, where exposure of subendothelial components may have occurred. The first layer of platelets is in contact with the thrombogenic surface (adhesion), whereas subsequent growth of the haemostatic plug depends on platelet–platelet interactions (aggregation). Both aspects of platelet function are influenced by von Willebrand factor binding to specific platelet membrane receptors as well as subendothelial structures, such as collagen.

1995 Cell adhesion and human disease. Wiley, Chichester (Ciba Foundation Symposium 189) p 35–50

The study of patients with von Willebrand's disease has been one of the main factors prompting the rapid progress in the understanding of the structure and function of von Willebrand factor. The knowledge acquired in this regard is proving important in delineating the complex processes responsible for platelet thrombus formation both during normal haemostasis and pathological thrombosis. What follows is a brief review of key concepts in these areas of research.

A review of experimental observations

Von Willebrand's disease, von Willebrand factor and platelet function

Von Willebrand's disease was first described in 1926 by Erik von Willebrand (von Willebrand 1926, 1931); it is the most common inherited human disorder of haemostasis, with a prevalence as high as 0.83% (Rodeghiero et al 1987) and is characterized by a complex haemostatic defect. Abnormal platelet function, expressed by prolonged bleeding time, is a consistent finding and may be accompanied by decreased factor VIII procoagulant activity. The pathogenesis of von Willebrand's disease is based on quantitative and/or qualitative abnormalities of von Willebrand factor, a large multimeric glycoprotein with two distinct biological roles: it mediates platelet adhesion and thrombus formation at sites of vascular injury, and it serves as the carrier for procoagulant factor VIII in circulating blood, where the two molecules are present as the factor VIII/von Willebrand factor complex (Weiss & Hoyer 1973, Weiss et al 1977). Mature von Willebrand factor has a typical multimeric structure and exists as a series of oligomers containing a variable number of subunits. Individual multimers range in mass from approximately 500 kDa to >10 000 kDa, the latter being the largest known for a soluble human plasma protein. The mature von Willebrand factor subunit contains 2050 amino acid residues and up to 22 carbohydrate side chains.

The gene encoding von Willebrand factor consists of about 180 kilobases and contains 52 exons (Mancuso et al 1989). It is located at the tip of the short arm of chromosome 12, region 12p12–12pter (Ginsburg et al 1985). A non-processed von Willebrand factor pseudogene has been identified on chromosome 22 (Mancuso et al 1991). The primary translation product predicted from the cloned von Willebrand factor cDNA is a 2813 residue precursor polypeptide referred to as prepro-von Willebrand factor (Bonthron et al 1986); it consists of a 22 residue signal peptide, an unusually large 741 residue propeptide and the mature subunit of 2050 residues (Fig. 1). The propeptide and mature subunit of von Willebrand factor are almost entirely composed of four types of repeating domains, designated A through D (Shelton-Inloes et al 1986). The von Willebrand factor propeptide is identical to a previously characterized protein, von Willebrand antigen II (Montgomery & Zimmerman 1978, Fay et al 1986). The normal synthesis of von Willebrand factor occurs in endothelial cells (Jaffe et al 1974) and megakaryocytes (Sporn et al 1985). Following translation of the mRNA, pro-von Willebrand factor undergoes extensive post-translational processing to produce multimeric von Willebrand factor (Wagner & Marder 1984, Wagner 1990).

Von Willebrand's disease exhibits significant phenotypic heterogeneity, depending on the particular subtype considered. Two main categories of patients can be distinguished on the basis of whether the main pathogenetic factor is a quantitative (type I and type III) or qualitative (type II) defect of von Willebrand factor.

Type I is the most common form of the disease, accounting for approximately 70% of all cases. It is inherited as an autosomal dominant, mild to moderately severe bleeding disorder. The disease, in this case, is a consequence of the presence of inadequate levels of von Willebrand factor in plasma, always accompanied by a parallel decrease in factor VIII procoagulant activity. Thus, although both von Willebrand factor and factor VIII are structurally and

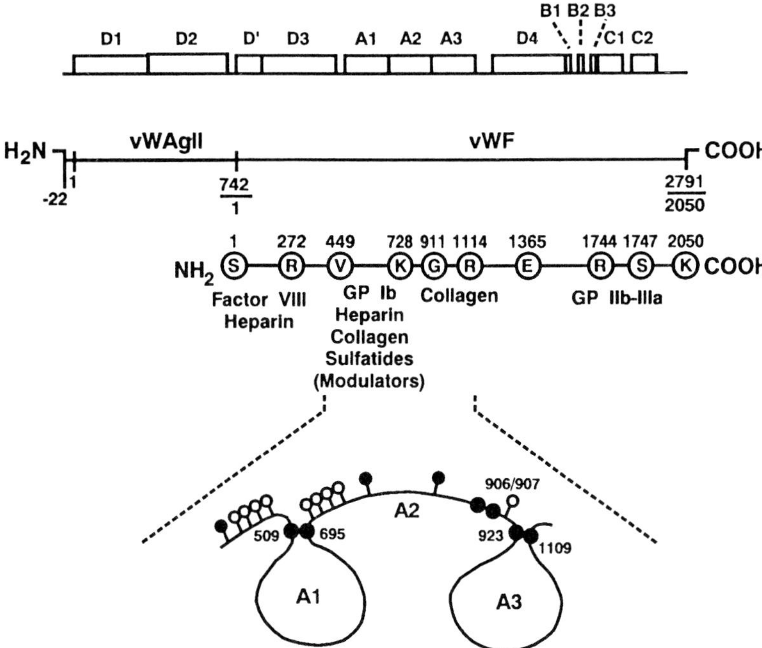

FIG. 1. Schematic structure of prepro-von Willebrand factor and functional domains in mature von Willebrand factor. The four types of cDNA repeated domains (A through D) are shown in the top line with their topographic arrangement relative to the linear amino acid sequence of prepro-von Willebrand factor. The latter (second line from top) is composed of the signal peptide beginning at position −22; 741 residues representing von Willebrand antigen II (vWAgII, now often referred to as the von Willebrand factor propeptide); and the 2050 residue mature von Willebrand factor subunit. Altogether, pro-von Willebrand factor contains 2791 residues. (Note: avoid confusion between pro-von Willebrand factor, the precursor of the mature molecule comprising vWAgII as well as the mature von Willebrand factor subunit, and the propeptide of von Willebrand factor, synonymous with vWAgII.) The third line from the top represents the main functional domains of mature von Willebrand factor identified to date, with an indication of the molecules they interact with. The boundaries of each domain are identified with the respective N- and C-terminal amino acid residue (one-letter notation); the first and last residue in the sequence of the mature von Willebrand factor subunit are also indicated. The scheme on the bottom represents in more detail the three type A domains of von Willebrand factor, with the large intrachain loops in A1 and A3. Cysteine residues are identified by black dots and the corresponding residue number in the sequence of the mature von Willebrand factor subunit. *O*-linked sugars are indicated by clear circles; *N*-linked sugars by filled circles. (Reprinted from Ruggeri & Ware 1993).

functionally intact in type I patients, the decreased concentration of von Willebrand factor, usually between 5% and 30% of normal, causes impaired function. Little is known, to date, about the molecular pathogenesis of type I von Willebrand's disease. Although it is generally accepted that most, if not all, cases must be the consequence of defects in the von Willebrand factor gene, no mutations responsible for any of the corresponding subtypes have yet been reported.

Type III von Willebrand's disease, transmitted as an autosomal recessive bleeding diathesis with severe to very severe manifestations, is the least common of all forms of von Willebrand's disease, with an incidence of approximately 1 in 1 000 000 subjects. Just as with type I, type III von Willebrand's disease is the consequence of a quantitative defect in von Willebrand factor, but the relative severity of the two forms is clearly different since the levels of plasma von Willebrand factor in the former type are usually undetectable even with very sensitive assays. Abnormalities of the von Willebrand factor gene have been detected in several families with type III von Willebrand's disease. These abnormalities vary from deletions of various size (from as small as 2.3 kb to the entire 178 kb gene) to single nonsense mutations (Ginsburg & Sadler 1993). Moreover, there is also evidence for the occurrence of *cis*-acting mutations affecting von Willebrand factor mRNA transcription, processing or stability (Nichols et al 1993).

Type II von Willebrand's disease comprises many different subtypes and is phenotypically very heterogeneous; common to all subtypes is the occurrence of qualitative abnormalities of von Willebrand factor, resulting, in most cases, in abnormal multimeric structure of the molecule. The disease, in these cases, is due to the existence of functional defects of von Willebrand factor that result in impaired platelet function even though the plasma concentration of von Willebrand factor (as well as factor VIII procoagulant) may be only modestly reduced or even normal. Single point mis-sense mutations of the von Willebrand factor gene have been identified in a number of patients, mainly with the two relatively frequent IIA and IIB subtypes of type II von Willebrand's disease (Ruggeri et al 1980, Ruggeri & Zimmerman 1980).

Platelets and shear forces

Platelet adhesion to thrombogenic surfaces. The first step in the response of platelets to vascular injury is their irreversible attachment to the altered surface. This stage is influenced by an essential aspect of the regulation of platelet function, namely the ability to recognize the site of lesion as different from the normal vessel wall. Two possible mechanisms can be considered as relevant in this regard; one, that substances (agonists) generated as a consequence of the lesion act on platelets in the proximity of the lesion and 'activate' them so that they can interact with adhesive molecules in blood or exposed at the site of injury; another, that circulating 'resting' platelets react with adhesive substrates exposed only where the vessel wall is altered. The regulation of platelet function would rest mainly, in the first instance, in their transition from 'resting' (inactive) to 'activated' and, in the second instance, in the limited exposure of efficient adhesive substrates to which 'resting' platelets can attach. In fact, a combination of both events is likely to occur during the formation of platelet thrombi. However, the response of 'resting' platelets to a local adhesive environment created by the lesion in the vessel wall probably represents the very initial response onto which the subsequent formation of a haemostatic plug is built. Indeed, it is possible that adhesion always precedes, and is necessary for, platelet activation.

Platelet aggregation. The process leading to a thrombus capable of arresting haemorrhage involves the interaction of platelets with one another. This occurs after the initial adhesion of platelets onto a thrombogenic substrate at the site of injury, and can actually be visualized as adhesion occurring onto platelets that are already adhering, or cohesion of platelets. Like adhesion, it is likely to involve both the recognition of activated adhering platelets as an appropriate adhesive substrate by 'resting' circulating platelets and the action of local agonists that 'activate' platelets. In this regard, it is relevant to consider that a significant proportion of the current concepts on the mechanisms of platelet aggregation is based on experimental models that study platelets in suspension, in the absence of a reactive surface. Nevertheless, during haemostasis *in vivo* aggregation is unlikely to occur independently of adhesion to a surface.

Blood viscosity, shear stress and platelet function. The complex series of events that determine the occurrence of a normal haemostatic response, including chemical reactions that involve cells, the vessel wall and molecules in solution, is influenced by the flow of blood. In a vessel, the velocity of blood near the wall is lower than towards the centre; this difference creates a shearing effect between adjacent layers of fluid moving at different speeds. Thus, 'shear' is the consequence of the relative parallel motion between adjacent fluid planes during flow; it is greatest near the vessel wall and decreases progressively towards the centre of the vessel. The local shear rate is expressed in cm/sec per cm, or the equivalent inverse seconds (s^{-1}). Fluid shear stress is the force per unit area that can be viewed as the underlying cause of the shearing motion of blood. The shear rate is directly proportional to the shear stress and inversely proportional to the viscosity of the fluid.

The role of platelets in haemostasis and thrombosis requires that they become irreversibly attached at a site of injury. This occurs against the tendency of flow to move adhering platelets with the layer of blood adjacent to the surface.

The force opposing a stable adhesion and aggregation is greater with increasing shear rate; consequently, it can be surmised that shear-dependent phenomena are of particular relevance in those districts of the vasculature where shear forces are greater, i.e. in arteries more than in veins and, particularly, in arterioles and capillaries. Moreover, in the presence of partial obstructions of the vessel lumen, as may be caused by atherosclerotic plaques and/or vasospasm, fluid shear stress may increase considerably above the average level of about 20 dynes/cm^2 calculated for the circulation of blood in normal vessels. The influence of shear-dependent phenomena, therefore, may be greater in pathological conditions predisposing to the occurrence of acute arterial occlusion than it is in the course of normal haemostasis.

Effect of shear stress on platelet aggregation: the role of von Willebrand factor

A good example of how shear forces can influence platelet functions is provided by comparison of the mechanisms involved in platelet aggregation under varying levels of shear. Typically, aggregation is evaluated with a device, the aggregometer, in which a suspension of platelets in plasma (platelet-rich plasma) is stirred with a bar rotating at the bottom of the cuvette containing the suspension. In this environment, shear stress varies depending on the position occupied by the platelets in the cuvette; on average, it is assumed to be low. Normal platelets do not aggregate when placed in an aggregometer and stirred unless an agonist is added, i.e. a substance that can 'activate' the platelets. Physiological agonists, those that may be present at a site of vascular injury, include thrombin, collagen and ADP; none is present in relevant amounts in circulating blood. In contrast, platelets can be exposed to different and homogeneous levels of shear stress in a device originally utilized to measure the viscosity of fluids, the cone-and-plate viscometer. In this instrument, the platelet-rich plasma is placed in the space between a stationary plate and a rotating cone. The distance between the two and the velocity of rotation of the cone determine the shear stress applied to the platelets. At levels above 60–80 dynes/cm^2, stable aggregation occurs without the need for exogenous agonists, suggesting that 'activation' may depend on mechanisms at least in part distinct from those responsible for aggregation measured in the aggregometer. The validity of the latter assumption is indeed demonstrated by the evidence that a specific platelet receptor, the membrane glycoprotein (GP) Ib/IX/V complex, and the adhesive protein that binds to it, von Willebrand factor, are necessary for aggregation induced by high shear (Peterson et al 1987, Ikeda et al 1991) but not for agonist-induced aggregation measured in the aggregometer.

Moreover, it is now established that the adhesive protein mediating aggregation through GPIIb/IIIa (also known as $\alpha_{IIb}\beta_3$) is different in a low shear environment (aggregometer) as opposed to a high shear one (cone-and-plate viscometer): fibrinogen is involved in the first instance, von Willebrand factor in the second. The role of von Willebrand factor in these processes appears to be most significant at high shear rates presumably as a consequence of its unique molecular architecture. Thus, it is possible that, under the effect of high shear forces, von Willebrand factor molecules take the shape of extended filaments, as seen with electron microscopy (Fowler et al 1985). The repeating subunit structure typical of these large multimers offers an array of interaction sites capable of binding in a multivalent manner to receptors on the platelet membrane, thereby increasing the number of contact points and the affinity (strength) of interaction (Federici et al 1989). As a result, the overall force linking platelets to the surface and/or to one another is increased, effectively opposing fluid shear stress. This interpretation of events explains why the role of von Willebrand factor is less relevant at lower shear rates, since other adhesive molecules may provide sufficient force of interaction to withstand opposing shear forces of lesser magnitude.

Hypotheses have been formulated to explain how shear stress can induce aggregation of platelets in suspension (Chow et al 1992, Ikeda et al 1993). It is believed that the binding of multimeric von Willebrand factor to GPIb with shear forces above 60–80 dynes/cm^2 causes a transmembrane flux of calcium ions increasing their intracellular concentration by two- to threefold. This results in activation of platelets, not dissimilar from that induced by 'agonists' in the aggregometer. Activation of platelets, in turn, confers to another receptor, GPIIb/IIIa, the ability to interact with soluble adhesive proteins and, thus, mediate platelet–platelet interaction (aggregation). Shear-induced aggregation depends on von Willebrand factor both for its initiation, to 'activate' platelets as other agonists can do in lower shear environments, and for its completion, to support platelet cohesion. This dual function is rather unique and is supported by two distinct sites in the molecule and two different receptors (Ikeda et al 1993, Savage et al 1992). The regulation of these processes depends on changes in the affinity of von Willebrand factor for GPIb, since soluble von Willebrand factor does not bind to the receptor unless an appropriate 'modulator' of the interaction is present. Two non-physiological modulators such as this have been used extensively—the antibiotic ristocetin (Howard & Firkin 1971, Scott et al 1991) and the snake protein botrocetin (Read et al 1978, Sugimoto et al 1991). At present there is no evidence that the mechanisms through which ristocetin and botrocetin induce binding of von Willebrand factor to GPIb are related to the pathophysiological regulation of this event *in vivo*; however, it is generally postulated that the two modulators, particularly botrocetin, may induce a conformational change in von Willebrand factor leading to the exposure of an otherwise hidden GPIb binding site. Thus, it may be hypothesized that a similar effect is achieved through the effects of mechanical shear forces acting either on the ligand (von Willebrand factor) or, more likely, on the receptor (GPIb). In any case, shear stress may be a relevant physiological modulator of von Willebrand factor binding to GPIb. Inhibition of the shear-induced von Willebrand factor–GPIb interaction results in the suppression of both the transmembrane calcium flux and aggregation; inhibition of von Willebrand factor binding to GPIIb/IIIa interferes only with aggregation, not with the calcium ion flux (Ikeda et al 1993). These findings establish the temporal sequence of events as 'activation' and 'cohesion', both von Willebrand factor dependent under high shear conditions.

Function of von Willebrand factor in platelet–surface and platelet–platelet interactions

The models of platelet aggregation are devised to measure platelet–platelet interactions occurring in the absence of a reactive surface. These conditions are different from those encountered during thrombus formation *in vivo*, where exposure of a thrombogenic surface to flowing blood represents the initial event leading to the local accumulation of platelets, eventually forming the haemostatic plug or the pathological occluding thrombus. The local adhesion of platelets recruited from the circulation represents the initial event of thrombus formation on a surface and is followed by accumulation of additional platelets onto the ones that adhere initially. While the latter process may be equivalent to the platelet–platelet interactions typical of aggregation, and may involve similar molecular mechanisms, the initial adhesion clearly involves interaction with immobilized, not soluble, adhesive proteins. This is a crucial difference, since platelet receptors exhibit remarkably distinct affinities and specificities for different ligands depending on whether the latter are immobilized or in solution. Moreover, the activation state of platelets also has a significant influence on the functional specificity of some receptors. Important examples are provided by the two main adhesion receptors, the GPIb/IX/V complex and the GPIIb/IIIa complex, and the adhesive ligands von Willebrand factor and fibrinogen.

When platelets are activated, the GPIIb/IIIa receptor is a promiscuous binding site, capable of interacting with at least four different ligands (fibrinogen, von Willebrand factor, fibronectin and vitronectin); in remarkable contrast, GPIIb/IIIa on non-activated platelets shows the ability to interact only with immobilized fibrinogen (Savage et al 1992, Savage & Ruggeri 1991). The GPIb component of the GPIb/IX/V complex, on the other hand, can interact with surface-bound von Willebrand factor even in the absence of shear stress, at variance with the situation discussed above for soluble von Willebrand factor. It is apparent, therefore, that GPIb and GPIIb/IIIa represent two pathways by which unstimulated platelets can attach to a thrombogenic surface that represents fibrinogen (fibrin) and/or von Willebrand factor exposed to flowing blood. There is, however, a fundamental difference between the processes initiated by the two different adhesive proteins. Irreversible adhesion to fibrinogen occurs even when platelets are metabolically inhibited. However, when activation inhibitors are present, no further platelet–platelet interaction and therefore no aggregation are observed. In contrast, when platelets are metabolically inhibited there is no irreversible adhesion onto a von Willebrand factor-coated surface, in spite of a normal initial contact with this adhesive protein.

Conclusions

On the basis of the experimental results presented above, it is possible to delineate an initial understanding of the mechanisms of platelet function that may allow thrombus propagation initiated and mediated by adhesive components at the site of vessel injury. Thus, the recognition of an appropriate surface-bound adhesive protein represents the first step in the process of thrombus formation; this platelet–surface contact must be followed by platelet spreading, necessary to make the initial contact irreversible (adhesion), and by activation, necessary to lead to cohesion of additional platelets into the forming thrombus. Spreading requires interaction of the surface-bound ligands with GPIIb/IIIa. In the case of fibrinogen (fibrin) it can proceed even if platelets are not fully activated since the receptor possesses the appropriate recognition specificity already expressed on resting platelets. In the case of von Willebrand factor, however, full platelet activation is necessary for GPIIb/IIIa to be able to recognize the ligand and mediate spreading. All these events can be demonstrated in the absence of flow and provide an example of how thrombus formation can be selectively initiated and amplified by specific adhesion receptors (Savage et al 1992). The latter clearly have a signalling function in addition to the one of mechanically supporting platelet attachment to a surface and to one another. Moreover, immobilized von Willebrand factor and fibrinogen (fibrin) act as proper platelet agonists, in addition to mediating adhesive events.

The shear stress in flowing blood has profound effects on the processes that lead to thrombus formation. As delineated above, the main one is that, with increasing shear rate, the pathways mediated by von Willebrand factor and GPIb acquire predominant relevance. In fact, thrombus formation onto a type I collagen surface is *independent* of von Willebrand factor up to a shear rate of 800 sec^{-1}, but becomes largely *dependent* on von Willebrand factor with increasing shear rate (Alevriadou et al 1993). As mentioned above, extreme values of shear stress never attained in the normal circulation (and, presumably, not of concern for the normal haemostatic processes) are reached under pathological conditions in stenosed atherosclerotic vessels. Thus, it is possible that selective pharmacological intervention aimed at blocking von Willebrand factor binding to GPIb (the interaction that initiates von Willebrand factor-dependent mechanisms of platelet adhesion and aggregation) may result in effective antithrombotic therapy with lesser haemorrhagic side effects. A detailed knowledge of how flow-related parameters influence the process of thrombus formation may, therefore, provide the means to understand fully the mechanism of normal haemostasis and to prevent pathological thrombosis.

References

Alevriadou BR, Moake JL, Turner NA et al 1993 Real-time analysis of shear-dependent thrombus formation and its blockade by inhibitors of von Willebrand factor binding to platelets. Blood 81:1263–1276

Bonthron D, Orr EC, Mitsock LM et al 1986 Nucleotide sequence of pre-pro-von Willebrand factor cDNA. Nucleic Acids Res 14:7125–7127

Chow TW, Hellums JD, Moake JL, Kroll MH 1992 Shear stress-induced von Willebrand factor binding to platelet glycoprotein Ib initiates calcium influx associated with aggregation. Blood 80:113–120

Fay PJ, Kawai Y, Wagner DD et al 1986 Propolypeptide of von Willebrand factor circulates in blood and is identical to von Willebrand antigen II. Science 232:995–998

Federici AB, Bader R, Pagani S, Colibretti ML, De Marco L, Mannucci PM 1989 Binding of von Willebrand factor (vWf) to glycoproteins (GP) Ib and IIb-IIIa complex: affinity is related to multimeric size. Br J Haematol 73:93–99

Fowler WE, Fretto LJ, Hamilton KK, Erickson HP, McKee PA 1985 Substructure of human von Willebrand factor. J Clin Invest 76:1491–1500

Ginsburg D, Sadler JE 1993 von Willebrand disease: a database of point mutations, insertions and deletions. Thromb Haemostasis 69:177–184

Ginsburg D, Handin RI, Bonthron DT et al 1985 Human von Willebrand factor (vWF): isolation of complementary DNA (cDNA) clones and chromosomal localization. Science 228:1401–1406

Howard MA, Firkin BG 1971 Ristocetin—a new tool in the investigation of platelet aggregation. Thromb Haemostasis 26:362–369

Ikeda Y, Handa M, Kawano K et al 1991 The role of von Willebrand factor and fibrinogen in platelet aggregation under varying shear stress. J Clin Invest 87:1234–1240

Ikeda Y, Handa M, Kamata T et al 1993 Transmembrane calcium influx associated with von Willebrand factor binding to GP Ib in the initiation of shear-induced platelet aggregation. Thromb Haemostasis 69:496–502

Jaffe EA, Hoyer LW, Nachman RL 1974 Synthesis of von Willebrand factor by cultured human endothelial cells. Proc Natl Acad Sci USA 71:1906–1909

Mancuso DJ, Tuley EA, Westfield LA et al 1989 Structure of the gene for human von Willebrand factor. J Biol Chem 264:19514–19527

Mancuso DJ, Tuley EA, Westfield LA et al 1991 Human von Willebrand factor gene and pseudogene: structural analysis and differentiation by polymerase chain reaction. Biochemistry 30:253–269

Montgomery RR, Zimmerman TS 1978 von Willebrand's disease antigen II: a new plasma and platelet antigen deficient in severe von Willebrand's disease. J Clin Invest 61:1498–1507

Nichols WC, Lyons SE, Harrison JS, Cody RL, Ginsburg D 1993 Severe von Willebrand disease due to a defect at the level of von Willebrand factor mRNA expression: detection by exonic PCR-restriction fragment length polymorphism analysis. Proc Natl Acad Sci USA 88:3857–3861

Peterson DM, Stathopoulos NA, Giorgio TD, Hellums JD, Moake JL 1987 Shear-induced platelet aggregation requires von Willebrand factor and platelet membrane glycoproteins Ib and IIb-IIIa. Blood 69:625–628

Read MS, Shermer RW, Brinkhous KM 1978 Venom coagglutinin: an activator of platelet aggregation dependent on von Willebrand factor. Proc Natl Acad Sci USA 75:4514–4518

Rodeghiero F, Castaman G, Dini E 1987 Epidemiological investigation of the prevalence of von Willebrand's disease. Blood 69:454–459

Ruggeri ZM, Ware J 1993 von Willebrand factor. FASEB (Fed Am Soc Exp Biol) J 7:308–316

Ruggeri ZM, Zimmerman TS 1980 Variant von Willebrand's disease: characterization of two subtypes by analysis of multimeric composition of factor VIII/von Willebrand factor in plasma and platelets. J Clin Invest 65:1318–1325

Ruggeri ZM, Pareti FI, Mannucci PM, Ciavarella N, Zimmerman TS 1980 Heightened interaction between platelets and Factor VIII/von Willebrand factor in a new subtype of von Willebrand's disease. N Engl J Med 302:1047–1051

Savage B, Ruggeri ZM 1991 Selective recognition of adhesive sites in surface-bound fibrinogen by GP IIb-IIIa on nonactivated platelets. J Biol Chem 266:11227–11233

Savage B, Shattil SJ, Ruggeri ZM 1992 Modulation of platelet function through adhesion receptors: a dual role for glycoprotein IIb-IIIa (integrin $\alpha_{IIB}\beta_3$) mediated by fibrinogen and glycoprotein Ib-von Willebrand factor. J Biol Chem 267:11300–11306

Scott JP, Montgomery RR, Retzinger GS 1991 Dimeric ristocetin flocculates proteins, binds to platelets and mediates von Willebrand factor-dependent agglutination of platelets. J Biol Chem 266:8149–8155

Shelton-Inloes BB, Titani K, Sadler JE 1986 cDNA sequences for human von Willebrand factor reveal five types of repeated domains and five possible protein sequence polymorphisms. Biochemistry 25:3164–3171

Sugimoto M, Mohri H, McClintock RA, Ruggeri ZM 1991 Identification of discontinuous von Willebrand factor sequences involved in complex formation with Botrocetin: a model for the regulation of von Willebrand factor binding to platelet glycoprotein Ib. J Biol Chem 266:18172–18178

Sporn LA, Chavin SI, Marder VJ, Wagner DD 1985 Biosynthesis of von Willebrand protein by human megakaryocytes. J Clin Invest 76:1102–1106

von Willebrand EA 1926 Hereditar pseudohemofili. Fin Lakaresallsk Handl 67:7–112

von Willebrand EA 1931 Ueber hereditaere pseudohaemophilie. Acta Med Scand 76: 521–550

Wagner DD 1990 Cell biology of von Willebrand factor. Annu Rev Cell Biol 6:217–246

Wagner DD, Marder VJ 1984 Biosynthesis of von Willebrand protein by human endothelial cells: processing steps and their intracellular localization. J Cell Biol 99: 2123–2130

Weiss HJ, Hoyer LW 1973 von Willebrand factor: dissociation from antihemophilic factor procoagulant activity. Science 182:1149–1151

Weiss HJ, Sussman II, Hoyer LW 1977 Stabilization of factor VIII in plasma by the von Willebrand factor. Studies on posttransfusion and dissociated factor VIII and in patients with von Willebrand's disease. J Clin Invest 60:390–404

DISCUSSION

Verrando: Are there many forms of von Willebrand's disease? Do you observe the same things in all cases?

Ruggeri: There are different subtypes of von Willebrand's disease. The most common form is one where you basically do not see, or cannot demonstrate, any functional abnormality of von Willebrand factor—all you see is a decreased concentration of it in the plasma. Some of these patients have a perfectly normal concentration of von Willebrand factor in their platelets. Most cases—I would say 60–70% of all patients—have a pure quantitative defect, and those patients bleed, so we know that von Willebrand factor is necessary above a certain level. Don't forget, these patients also have a concurrent decrease of factor VIII, which may contribute to bleeding. Then there is a whole series of subtypes with very specific phenotypes. In terms of single point mutations, the only functions that have been shown to be affected are the following: in some patients von Willebrand factor loses the ability to bind to factor VIII to form the complex. These patients have a haemophilia-like syndrome, because their only defect is a low level of factor VIII in the blood—their von Willebrand factor is normal and their platelet function is normal. In patients who have mutations in the A1 domain of von Willebrand factor, the molecule interacts abnormally with GPIb and they bleed. There is no known mutation in patients that affects binding of von Willebrand factor to collagen or GPIIb/IIIa. Under all conditions where there is an abnormal function of von Willebrand factor, there is a platelet defect that is reflected mainly in loss of adhesion at high shear.

Etzioni: Do you think that thrombotic thrombocytopenic purpura (TTP) is the opposite of von Willebrand's disease?

Ruggeri: TTP is a very complex disease characterized by the presence of platelet thrombi with minimal (if any) fibrin participation, in the arterial microcirculation. This results in organ damage owing to the blockage of blood flow—with mainly neurological and kidney symptoms. Because just platelets are involved, we don't know what causes this aggregation *in vivo*. It is probably the only example in real life where pure platelet aggregation occurs. I don't know whether it's fair to suggest that it is the opposite of von Willebrand's disease: it's certainly different because patients might have pathogenesis owing to enhanced interaction of von Willebrand factor with GPIb.

Etzioni: There are some reports about increased levels of von Willebrand factor in TTP (Moake 1988).

Ruggeri: Yes, but von Willebrand factor is a reactive protein that might increase as a secondary phenomenon. There's no evidence that just increasing the concentration of von Willebrand factor in blood leads, by itself, to platelet aggregation. It's actually quite clear that it doesn't.

Humphries: Is the dependence on shear rate affected by ligand density on the substrates, for example, or can you force collagen to support adhesion at low flow rate by increasing its density?

Ruggeri: No. It appears that substrate density has no (or minimal) effect on the ability of collagen to support adhesion (or not) at a given wall shear rate.

Pober: In *in vitro* flow chambers, why do larger forms of von Willebrand factor appear to bind platelets better than smaller forms?

Ruggeri: There has been no experimental demonstration but, intuitively, the answer is you need a multivalent ligand to perform this particular function. When you think about the surface, the idea of a multivalent ligand is problematic, because if you take a monovalent ligand and put it on the surface at a sufficient density, it could mimic a multivalent ligand very well. What I think you need is a particular orientation of von Willebrand factor and a certain density of GPIb binding sites, because I believe that cross-linking of GPIb is crucial for activation. All these reactions require active responses.

Wagner: Moake et al (1986) showed that in TTP the very large von Willebrand factor multimers may be aggregating platelets at high shear. Don't you believe this any more?

Ruggeri: In that regard you would have to believe that either von Willebrand factor is somehow changed, or that there is some other molecule that contributes to increasing the affinity of von Willebrand factor for GPIb. In fact, we know that soluble von Willebrand factor does not bind to GPIb; we can increase the concentration of normal von Willebrand factor as much as we want, but we cannot measure an interaction. We also know that there are single point mutations in von Willebrand's disease (type IIB) where there is an increased affinity for GPIb. In these patients, soluble von Willebrand factor does bind to GPIb, but they don't have TTP; on the contrary, they bleed because soluble von Willebrand factor binds to GPIb, blocking the receptor; then, when they have a vascular lesion, the GPIb receptor can no longer support adhesion by interacting with von Willebrand factor in the vessel wall. Thus, in a situation where the affinity of von Willebrand factor for GPIb is clearly increased, the pathological consequences are not reminiscent of TTP: in fact, it is quite the opposite—the patients bleed.

Shaltiel: In the cases where there is a high affinity for GPIb with the monomer, is there an effect on the multimerization?

Ruggeri: Yes, the large multimers are decreased in number because they're bound to the platelet surface in circulating blood and, therefore, are effectively removed. These patients often also have thrombocytopenia, because of the binding of von Willebrand factor and the formation of small aggregates in the circulation. This is actually similar to TTP, but you never see thrombosis in these patients. Obviously, in TTP you may have pathology at the level of endothelial cells, which could be very important for localization of small thrombi.

Wagner: Soluble von Willebrand factor doesn't bind to GPIb. What is the conformational change that von Willebrand factor undergoes when it binds to the substratum which makes it adhesive?

Ruggeri: This has never been addressed experimentally in a definitive way. The idea that conformational changes in von Willebrand factor might be relevant in changing its affinity for GPIb is obviously a possibility, but it has never been proven. By itself, the binding of a protein to a surface can change a lot of things in that protein. For example, fibrinogen bound to a surface (unlike soluble fibrinogen) becomes a ligand for non-activated GPIIb/IIIa. I don't know whether there is a specific conformational change of von Willebrand factor that depends on its interaction with a specific component of the subendothelium. Most likely, shear forces can affect GPIb, so the other possibility is that the receptor could also undergo conformational and functional regulation.

Hynes: Some of your results clearly show that you're exposing the binding site in the von Willebrand factor, in the sense that your recombinant A1 domain does bind. Perhaps, if you treat with botrocetin, you expose something. So it has to be in some sense a conformational change.

Ruggeri: One has to be a little bit cautious in interpreting these results. For example, there are ways (and you mentioned one) to induce the von Willebrand factor to bind to GPIb; however, these mechanisms may not reflect what occurs *in vivo*. Botrocetin forms a complex with the A1 domain and, in this regard, it's a very appealing model. The problem is that it is not at all clear that it isn't botrocetin itself that, once it's bound to von Willebrand factor, participates in the binding to GPIb, rather than simply inducing a conformational change in von Willebrand factor. The reason for saying this is that there are molecules that are very homologous to botrocetin that will bind to GPIb, even without von Willebrand factor. In a sense, therefore, it could actually be that von Willebrand factor modulates botrocetin binding to GPIb. To address your question, I believe that even without shear effects, GPIb will bind to surface-bound von Willebrand factor. The evidence comes from our flow studies. Platelets adhere to surface-bound purified von Willebrand factor at any shear, even under static conditions. What is intriguing about this is that on collagen type I, there is minimal adhesion at low shear. Shear seems to be necessary for the interaction between von Willebrand factor and collagen to occur, and that's probably the most relevant observation as far as von Willebrand factor platelet function is concerned.

Garrod: A similar sort of phenomenon occurs with fibrinogen (Andrieux et al 1989); is anything known about that interaction?

Ruggeri: With regard to adhesion of non-activated platelets to fibrinogen, the γ-chain dodecapeptide is necessary but not sufficient to mediate the adhesion. We have an antibody that maps to the E-domain of fibrinogen but we know it is not against the RGDF sequence. This antibody selectively inhibits adhesion of platelets, mediated by non-activated GPIIb/IIIa to fibrinogen. We are presently isolating this domain and we're trying to figure out what's going on. Purified fragment D of fibrinogen, which contains the γ-chain dodecapeptide sequence in a monovalent conformation, does not support this adhesion. It supports adhesion quite well after activation of GPIIb/IIIa, but not with non-activated GPIIb/IIIa. So we think there are (at least) two domains, both of which are necessary but neither one of which is sufficient to support this interaction.

Hynes: You mentioned two activated states of the GPIIb/IIIa integrin receptor; do you think there are just two or do you think there might be more than that? You mentioned the extremes: the non-activated state, which will bind surface bound fibrinogen, and the fully activated state. There's some evidence that fits best for an intermediate state that will activate part way (Kieffer & Phillips 1990).

Ruggeri: There is a relatively simple experiment you can do concerning this, the interpretation of which seems easy but could be complex. Treatment of platelets with prostaglandin E_1 blocks full activation (if you define full activation of platelets as that particular state where α-granule content has been released and GPIIb/IIIa has been turned into a receptor that will bind soluble adhesive proteins); yet prostaglandin E_1-treated platelets under shear will irreversibly adhere and spread on fibrinogen, reaching what I believe to be an intermediate state of activation.

Stanley: What are the prospects of gene therapy for this disease?

Ruggeri: The prospect is that everybody is thinking about haemophilia and very few people are thinking about von Willebrand's disease.

Stanley: Is there a biological reason for that?

Ruggeri: The size of the gene might scare people, but, of course, the gene for haemophilia is not small. The vast majority of cases of von Willebrand's disease are mild, so the need is not as great as it is for haemophilia. Undoubtedly, this is something that some day will become easily available.

Stanley: You seem to imply that there's a quantitative defect in some patients.

Ruggeri: These are the patients who have very mild disease; they are the patients in whom you can already achieve pretty good therapeutic results with non-transfusional methods, because there are substances that will induce release of von Willebrand factor from the endothelium. Von Willebrand's disease is very common; its prevalence is about 0.8% of the population. But 70% of the cases are very mild and respond very well to non-transfusional therapy. That's great because everybody prefers not to use blood products because of the obvious risks associated with them. Severe cases do exist where patients lack von Willebrand factor; therefore, their factor VIII is very low and they have haemophilia-like bleeding. These are the patients who would benefit from more definitive therapeutic approaches.

Stanley: Where is the von Willebrand factor normally synthesized and secreted?

Ruggeri: It is synthesized by endothelial cells and megakaryocytes, but it is secreted only by endothelial cells. To the best of my knowledge the megakaryocytic product is only stored in α-granules and is not normally released until platelets are activated, whereas the constitutive secretion and regulated secretion (what we see in the circulation) comes from endothelial cells.

Wagner: We have cultured megakaryocytes from a leukaemic patient and they did secrete von Willebrand factor constitutively (Sporn et al 1985). Therefore, some von Willebrand factor in plasma could also be from this source.

Ruggeri: There was a study on bone marrow transplantation in pigs with severe von Willebrand's disease. In these animals, who after the graft had normal platelets containing von Willebrand factor, the plasma levels of circulating von Willebrand factor showed minimal changes, implying that secretion from the megakaryocytes is negligible.

Wagner: Indeed, they replenished very little plasma von Willebrand factor.

Sonnenberg: What's the role of the other integrins that are expressed on platelets: $\alpha_2\beta_1$, $\alpha_5\beta_1$ and $\alpha_6\beta_1$?

Ruggeri: I have no doubt that $\alpha_2\beta_1$ (the collagen receptor) has a very important role in supporting platelet adhesion and thrombus formation.

Sonnenberg: At low shear rates?

Ruggeri: I think it's relative. The collagen receptor is also synergistic with the GPIb/IX/V complex leading to activation of platelets that attach to von Willebrand factor, which in turn is complexed to collagen. But just to give you an idea of how complex the situation might be, it could depend on the type of collagen, because collagen type VI will bind von Willebrand factor when it's put on the surface, as will collagen type I or III; but collagen type VI does not support adhesion at high shear rates.

Hynes: Have you looked at the blood from α_2-integrin-deficient patients?

Ruggeri: No, but it's very clear that those patients bleed, and that fits with the idea that platelet adhesion at high shear is a function of von Willebrand factor and collagen in synergism.

Wagner: Platelets bind to fibrin: is this mediated through von Willebrand factor? Is there a fibrin receptor on platelets, or is the fibrinogen receptor the same as the fibrin receptor?

Ruggeri: Under experimental conditions, platelet adhesion to either fibrinogen or fibrin is completely inhibited by anti-GPIIb/IIIa antibodies. This suggests that a receptor common to fibrinogen and fibrin is involved in the interaction, but does not exclude the participation of other adhesive proteins. More experimental work needs to be done to clarify these complex issues.

References

Andrieux A, Hudry-Clergeon G, Ryckewaert J-J et al 1989 Amino acid sequences in fibrinogen mediating its interaction with its platelet receptor, GPIIb/IIIa. J Biol Chem 264:9258–9265

Kieffer N, Phillips DR 1990 Platelet membrane glycoproteins: functions in cellular interactions. Annu Rev Cell Biol 6:329–357

Moake JL 1988 von Willebrand factor and the pathophysiology of thrombocytopenia: from human studies to a new animal model. Lab Invest 59:415–417

Moake JL, Turner NA, Stathopoulos NA, Molasco LH, Hellums JD 1986 Involvement of large plasma von Willebrand factor (vWf) multimers and unusually large vWf forms derived from endothelial cells in shear stress-induced platelet aggregation. J Clin Invest 78:1456–1461

Sporn LA, Chavin SI, Marder VJ, Wagner DD 1985 Biosynthesis of von Willebrand protein by human megakaryocytes. J Clin Invest 76:1102–1106

Leukocyte adhesion deficiency (LAD) II

A. Etzioni*, L. M. Phillips†, J. C. Paulson† and J. M. Harlan‡

*Department of Pediatrics, Rambam Medical Center, B. Rappaport Medical School, Haifa, Israel, †Cytel Corporation, San Diego, CA and ‡Division of Hematology, University of Washington, Seattle, WA, USA

Abstract. The occurrence of recurrent bacterial infections, neutrophil motility dysfunction and normal expression of β_2 integrins (CD18) in two unrelated children suggested an as yet undescribed adhesion deficiency. The fact that both children exhibited the rare Bombay blood group and were Lewis negative, each involving carbohydrates with different fucose linkages, suggested a possible defect in the fucose-containing ligand for E- and P-selectin, sialyl Lewis X (SLe^x). Using a monoclonal anti-SLe^x antibody, we did not detect expression of SLe^x on the neutrophils of the patients. Adhesion of neutrophils to endothelial cells activated with interleukin-1β or histamine was markedly decreased ($<5\%$ of control). The observation that the neutrophils did not bind to recombinant E-selectin and purified P-selectin confirmed the SLe^x deficiency as the basis for adhesion deficiency. Using several *in vivo* techniques, we were able to show that neutrophil rolling, the first step in their adhesion, is markedly decreased, and therefore neutrophil emigration through the endothelium and arrival at site of inflammation is significantly diminished (1–2% of normal). Low binding of fucose-specific lectins to the patients' B lymphocytes transformed with Epstein–Barr virus was observed, while the binding of mannose-specific lectins was normal, providing further evidence for a general fucose deficiency as the primary defect. The existence of the patients and their deficiency emphasizes the essential role of the endothelial cell selectins and their ligand, SLe^x, in recruitment of neutrophils to sites of infection.

1995 Cell adhesion and human disease. Wiley, Chichester (Ciba Foundation Symposium 189) p 51–62

'It is in her moments of abnormality that Nature reveals her secrets' (Goethe).

Neutrophils are the front-line defence of mammals against most microbial pathogens. They provide a rapid, relatively non-specific defence mechanism, after which a more long-lasting antigen-specific response is established by T and B lymphocytes (Etzioni & Douglas 1993). The process of neutrophil localization is dynamic and involves multiple steps (Springer 1994). These steps must be precisely orchestrated to ensure a rapid response to isolate and destroy the invading pathogens. Neutrophil interactions with vascular endothelial cells are of central importance in guiding the acute inflammatory response and are mediated by three adhesion molecule families: the integrins, the immunoglobulin superfamily and the selectins (Etzioni 1994).

The first step in neutrophil recruitment is the rolling phase. In this step, adhesion is loose, transient and reversible, but is a prerequisite for the next step. This first step is mediated by the selectins and their ligands on neutrophils. In the second step, neutrophil activation occurs with activation of the β_2 integrin molecules (CD18). In the third and final step, firm adhesion and migration of the cells occur; these are dependent on integrins and immunoglobulin superfamily molecules (Bevilacqua 1993).

Important insights into leukocyte emigration and its molecular basis have come from studies of the leukocyte adhesion deficiency (LAD) syndromes I and II, two rare inheritable disorders caused by deficiency of the β_2 integrins (LAD I) (Harlan 1993) and sialyl Lewis X (SLe^x), the ligand for selectin (LAD II) (Etzioni et al 1992). This review will focus mainly on our current knowledge of LAD II, comparing it to LAD I.

The Clinical Picture

LAD II was first described in two unrelated boys, each the offspring of consanguineous parents. They were born after uneventful pregnancies with normal height and weight. No delay in the separation of the umbilical cord was observed. Both children have severe mental retardation, short stature, a distinctive facial appearance and the rare Bombay (hh) blood phenotype (Frydman et al 1992, Etzioni et al 1993). From early life they have both suffered from recurrent episodes of bacterial infection (mainly pneumonia, periodontitis, otitis media and localized cellulitis without pus formation). These are not life-threatening events and are usually treated in the outpatient clinic. In the last three years, the frequency of infections has decreased and they have been taken off prophylactic antibiotics. From the first days of their life, extreme neutrophilia was noted. During infections their neutrophil count goes up to 100 000/ml and ranges between 20–35 000/ml when they are free of infections. The basic defect in LAD II is a general defect in fucose metabolism, causing the absence of the H antigen on erythrocytes (Bombay phenotype), the SLe^x antigen on neutrophils, and also of other fucosylated molecules. The syndrome is inherited in an autosomal recessive mode and, recently, prenatal diagnosis of LAD II was made by revealing the Bombay blood phenotype in a 21 week female fetus in one of the families (Frydman et al 1994).

The severity of the clinical symptoms is less than that observed in the classical form of LAD I, in which a high mortality rate is observed in early childhood. LAD II is more compatible with the moderate type of LAD I (Table 1).

TABLE 1 Clinical and laboratory features of leukocyte adhesion deficiencies (LAD) I and II

	LAD I	*LAD II*
Clinical Manifestation		
Recurrent severe infection	+ + +	+
Neutrophilia	+ + +	+ + +
Gingivitis	+ +	+ +
Skin infection	+ +	+ +
Delay in separation of umbilical cord	+ + +	Normal
Developmental abnormalities	Normal	+ + +
Laboratory finding		
β_2 integrin expression	↓↓↓	Normal
SLe^x expression	Normal	↓↓↓
Neutrophil motility	↓↓↓	↓↓↓
Neutrophil adherence	↓↓↓	↓↓↓
Neutrophil rolling	Normal	↓↓↓
Opsonin-induced phagocyte activity	↓	Normal
T and B cell function	↓	Normal

Degree of clinical manifestation indicated by +, + + or + + +; in the laboratory findings, the reduction in expression/activity is indicated by ↓, ↓↓, or ↓↓↓.

Neutrophil Studies

The clinical picture (skin infection with no pus formation, pneumonia and peridontitis) and the very high neutrophil count, suggest that a neutrophil defect is the cause of this syndrome. While the opsonophagocytic activity of the patients' neutrophils was normal, we found a marked defect in their motility (Etzioni et al 1992): both random migration and intentional migration towards chemotactic factors were markedly decreased (10% of normal). Furthermore, the homotypic aggregation of these neutrophils was found to be absent (Etzioni et al 1994a). These results confirmed the problem as a defect in the ability of neutrophils to adhere to surfaces; therefore we looked at a defect in adhesion molecules. Indeed, while the neutrophils of the patients and their parents exhibit normal levels of expression of the integrin subunits, LAD II neutrophils are deficient in expression of the SLe^x antigen (Etzioni et al 1992) (Fig. 1).

Other fucosylating antigens, such as Lewis X, are also lacking on the patients' cells. Normal expression of these antigens is found on neutrophils from their parents (Phillips et al 1994).

To look at the significance of this defect, we did both *in vitro* and *in vivo* experiments. Neutrophils isolated from peripheral blood of LAD II patients

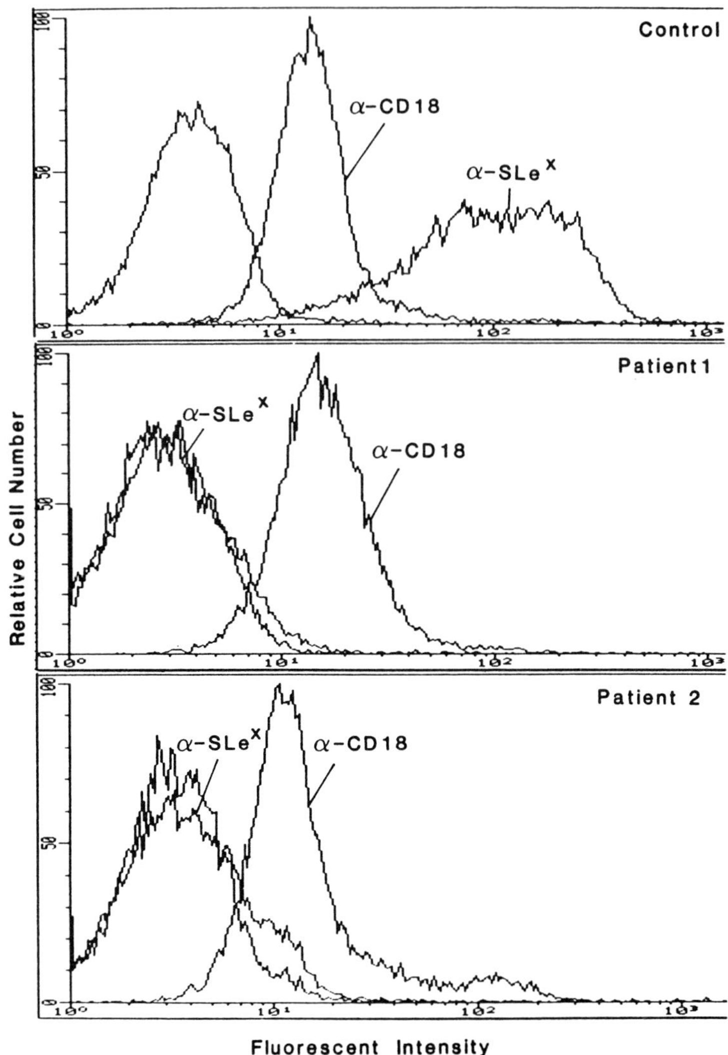

FIG. 1. Expression of SLe^x and CD18 (β_2 integrins) on neutrophils from patients with LAD II and control. (Reproduced with permission from Etzioni et al 1992.)

were examined *in vitro* for their ability to bind to purified P-selectin derived from platelets and purified recombinant E-selectin.

We found that these neutrophils did not bind to either selectin (Phillips et al 1994). When we used activated human umbilical vein endothelial cells, only activation with phorbol myristate acetate (PMA) (inducer of β_2 integrin) promoted adhesion (Etzioni et al 1993) (Fig. 2).

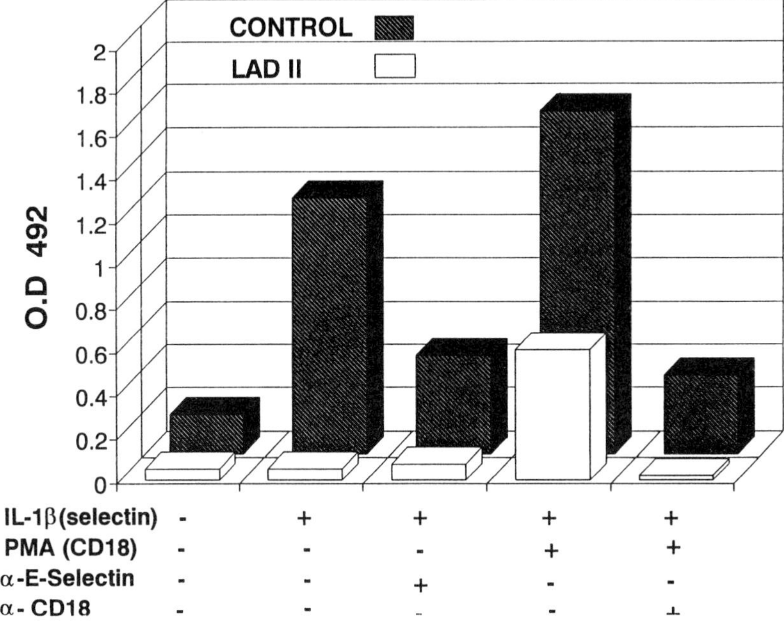

FIG. 2. Binding to human umbilical vein endothelial cells by LAD II and control neutrophils, as measured by ELISA, in the presence of various activators and blockers of adhesion. PMA, phorbol myristate acetate (inducer of CD18 [β_2 integrin]); IL-1β, interleukin 1β (inducer of E-selectin); α-E-selectin, anti-E-selectin antibodies (blocker of E-selectin); α-CD18, anti-CD18 antibodies (blocker of CD18).

Rolling, the first step in neutrophil recruitment to site of inflammation, is thought to be mediated by the selectins and their ligand, SLe^x. To determine the relative ability of patients' neutrophils to roll, we studied the *in vivo* behaviour of fluorescence-labelled neutrophils during their passage through inflamed venules in rabbit mesenteries using intravital microscopy (von Andrian et al 1991). The rolling fraction of control neutrophils in this assay is around 30%. Similar results were observed when LAD I neutrophils were studied. In contrast to control and LAD I cells, LAD II neutrophils rolled poorly: rolling fraction was only 5% (Fig. 3), and most of the cells that did interact with endothelial cells had a higher rolling velocity, rolled only over short distances and frequently detached from the vessel wall (von Andrian et al 1993). This remarkable inability to react to an inflammatory stimulus appears to be dependent on the presence of intravascular shear force. In an additional experiment, the mesenteric blood flow was stopped and LAD II cells were injected into the unperfused microvasculature. When blood flow was later restored, numerous LAD II cells had become stuck

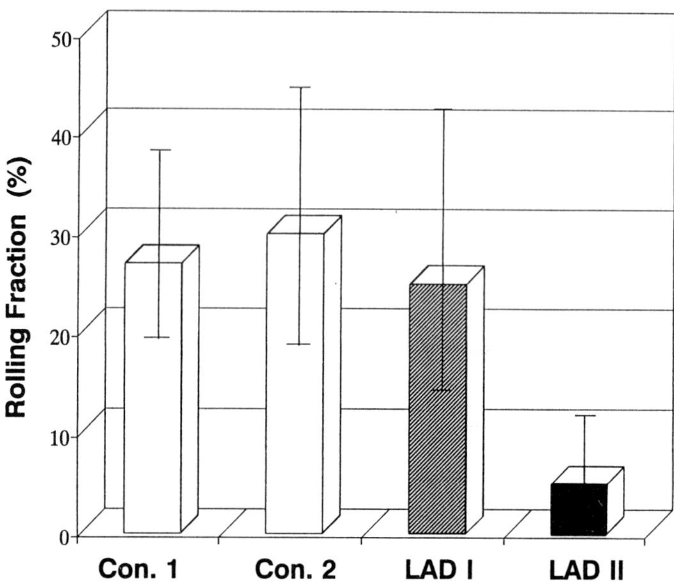

FIG. 3. Rolling of neutrophils from controls and LAD I and LAD II patients, as measured by the intravital microscopy assay. Error bars represent standard deviation of 10 venules analysed.

and were not detached by the flowing blood. These experiments indicate that the immune defect in LAD II is the result of a shear-dependent inability of neutrophils to roll and slow down in inflamed venules and not due to a dysfunction of shear-independent later steps in the intravascular adhesion process.

Following these results, we went on to examine the *in vivo* chemotaxis of the patients' neutrophils. The response to cutaneous inflammation was assessed by both skin chamber and skin window techniques. Neutrophil emigration was markedly diminished in both tests, the values being approximately 1.5% and 6% of normal in the skin window and skin chamber tests, respectively (Price et al 1994). Monocyte migration to the skin window site was reduced to a similar degree. Neutrophils from patients with LAD I showed the same magnitude of defect in these tests.

Adhesion molecules were also found to participate in the immune reaction at the level of T lymphocyte function (Springer 1990). Indeed, several lymphocyte functions were defective in LAD I (Fisher et al 1988). Looking at the immune function in LAD II, we detected normal proportions and numbers of subpopulations of T lymphocytes. Their proliferative response to various mitogens and their natural killing activity, as well as immunoglobulin levels, were all within the normal range (Etzioni et al 1994b). Furthermore, antibody production after immunization with the bacteriophage OX174, was found to be normal, with the normal switch from IgM to IgG. This is in contrast to patients with LAD I. Still, skin tests to various antigens were all negative. This anergy is most probably a result of the absence of the cutaneous lymphocyte antigen from the lymphocyte (Etzioni et al 1994b).

Conclusion

A new adhesion molecule defect, LAD II, has enabled us to unravel the role of adhesion molecules in host defence mechanisms. The general defect in fucose metabolism, which is the basic abnormality in this syndrome, is responsible for the absence of various fucosylated molecules, including SLe^x, the ligand for the selectins. The cause of short stature and mental retardation in the two LAD II patients is still unknown. Both children were born normally and no intrauterine growth retardation was observed. It is not clear whether these abnormalities are related to defects in adhesion or to a yet unknown function of fucosylated molecules.

It has recently been proposed that adhesive interactions between leukocytes and endothelial cells involve multiple sequential steps (Butcher 1991, Springer 1994). The final solution to the role of each step can come only from those rare experiments of Nature where a single defect is found and can be related to the clinical picture. The recent development of techniques for knocking out specific genes has enabled us to learn more about the role of various molecules. A P-selectin-deficient mouse model was developed by Mayadas et al (1993) with features quite similar to those observed in LAD II. The crucial *in vivo* role of the integrin family was established when LAD I was discovered. Patients with this disorder suffer from severe bacterial infections and, in its most severe form, death in early childhood will occur if bone marrow transplantation is not performed.

The newly described LAD II syndrome clarifies the role of the selectin family and their ligand, SLe^x. It is clear from our *in vitro* and *in vivo* studies that this family of adhesion molecules is essential for the first step in neutrophil emigration through the blood vessel, the rolling phase. Clincally, patients with LAD II suffer from a less severe type of disease resembling the moderate phenotype of LAD I. This is due, at least in part, to the observed normal T and B lymphocyte function in LAD II as opposed to LAD I.

References

Bevilacqua MP 1993 Endothelial—leukocyte adhesion molecules. Annu Rev Immunol 11:767–804

Butcher EC 1991 Leukocyte—endothelial cell recognition: three (or more) steps to specificity and diversity. Cell 67:1033–1036

Etzioni A 1994 Adhesion molecules in host defense. Clin Diagr Lab Immunol 1:1–4

Etzioni A, Frydman M, Pollack S et al 1992 Severe recurrent infections due to a novel adhesion molecule defect. N Engl J Med 327: 1789–1792
Etzioni A, Harlan JM, Pollack S, Phillips LM, Gershoni-Baruch R, Paulson JC 1993 Leukocyte adhesion deficiency (LAD) II: a new adhesion defect due to absence of sialyl lewis x, the ligand for selectins. Immunodeficiency 4:307–308
Etzioni A, Douglas SD 1993 Microbial phagocytosis and killing in host defense. In: Spirer Z, Roifman CM (eds) Pediatric Immunology. Karger, Basle, p 17–27
Etzioni A, Gershoni-Baruch R, Harlan JM 1994a Sialyl lewis x and neutrophil aggregation. Blood 83:876–877
Etzioni A, Kuypers T, Pollack S, Pals S 1994b Lymphocyte function in LAD II. In preparation
Fisher A, Lisowska-Grospierre B, Anderson DC, Springer TA 1988 Leukocyte adhesion deficiency: molecular basis and functional consequences. Immunodefic Rev 1:39–54
Frydman M, Etzioni A, Eidlitz-Markus T et al 1992 Rambam-Hasharon syndrome of psychomotor retardation, short stature, defective neutrophil motility, and Bombay phenotype. Am J Med Genet 44:297–302
Frydman M, Verdimon D, Shaler E, Orlin GB 1994 Prenatal diagnosis of LAD II syndrome. Prenatal Diagnosis, in press
Harlan JM 1993 Leukocyte adhesion deficiency syndrome: insight into the molecular basis of leukocyte emigration. Clin Immunol Immunopathol 69:S16–S24
Mayadas TN, Johnson RC, Rayburn H, Hynes RO, Wagner DD 1993 Leukocyte rolling and extravasation are severely compromised in P selectin-deficient mice. Cell 74:541–554
Phillips ML, Schwartz BR, Etzioni A et al 1994 Neutrophil adhesion in leukocyte adhesion deficiency syndrome type 2 (LAD II). J Clin Invest, submitted
Price TH, Harlan JM, Ochs HD, Gershoni-Baruch R, Etzioni A 1994 In vivo neutrophil and lymphocyte function studies in a patient with leukocyte adhesion deficiency type II. Blood 84:1635–1639
Springer TA 1990 Adhesion receptors of the immune system. Nature 346:425–434
Springer TA 1994 Traffic signals for lymphocyte recirculation and leukocyte emigration: the multiple paradigm. Cell 76:301–314
von Andrian UH, Chambers JD, McEvoy LM, Bargatze RF, Arfors KE, Butcher EC 1991 Two-step model of leukocyte–endothelial cell interaction in inflammation: distinct roles for LECAM-1 and the leukocyte β_2 integrins in vivo. Proc Natl Acad Sci USA 88:7538–7542
von Andrian UH, Berger EM, Ramezani L et al 1993 In vivo behavior of neutrophils from two patients with distinct inherited leukocyte adhesion deficiency syndromes. J Clin Invest 91:2893–2897

DISCUSSION

Hynes: We've discussed before the possibility of treating LAD II patients by feeding them fucose. The problem with this is that it generates the H blood group. Is there a way that you could inhibit the specific transferase that makes the H blood group and then feed them fucose?

Etzioni: Not that I know of.

Stanley: If you could get H expressed in the fetus, you might avoid the immune problem.

Etzioni: When the LAD II patients were born, they were completely normal regarding their height, weight and head circumference. The problems of mental retardation and short stature only appeared later on. Initially, we thought that during fetal life fucose might be transferred from the mother to the fetus. In cows' milk there is usually very little fucose, but in human milk there's quite a lot. Consequently, we assumed that LAD II patients wouldn't have problems for the first several months of their lives while they were breast fed. But we went back to our patient and found that in his first week of life he already had a severe bacterial infection with a very high leukocyte count (100 000/mm^3). Now we know for sure that LAD II patients don't have fucose in fetal life because subsequently we have had an aborted LAD II female who was found to be Bombay phenotype, so there isn't enough transfer of fucose from the mother to the child to compensate for the deficiency.

Rothlein: We envision that selectins interact with SLe^x and that this is important for leukocyte–endothelial cell interactions. I'm not sure why you're getting results *in vitro*, where neutrophils don't spread on glass or don't migrate through agar (Etzioni et al 1994), because the neutrophil lacks SLe^x.

Etzioni: You're not the only one who doesn't understand them! It is not clear why the lack of SLe^x is a factor on leukocyte migration through agar or on glass.

Haskard: Could the reduced spreading on glass of the neutrophils from LAD II patients be due to contaminating platelets in the normal neutrophil preparation, which in some way mediate the spreading response? Failure to bind platelets might then explain the reduced spreading of neutrophils from patients with LAD II.

Etzioni: We have repeated this experiment several times with controls.

Haskard: It's very difficult to prepare pure populations of leukocytes without contaminating platelets.

Wagner: These leukocytes lacking the carbohydrate ligand for P-selectin should not interact with platelets.

Rothlein: Has anybody tried to repeat these spreading and migration experiments with anti-SLe^x antibodies using normal leukocytes?

Etzioni: Not that I know of.

Elices: Are the other defects that you find in these patients, such as their mental retardation and their short stature, related in any way to SLe^x deficiency?

Etzioni: Santos-Benito et al (1992) showed that SLe^x is important for normal brain development. When the LAD II children were born, they didn't have microcephaly as you see in many other cases where there is a defect in brain development. So it seems more reasonable to say that SLe^x is important postnatally.

Ley: Can you analyse the SLe^x or structural analogues from the children to find out whether the fucose component is replaced by another sugar?

Etzioni: We haven't thought of trying that.

Ley: It seems to me that if the fucose sugar was exchanged for another sugar, this could be crucial for the adhesion properties.

Hynes: Does it often happen that when you are missing an enzyme some other enzyme comes in?

Ley: Yes, it could happen easily and could lead to novel SLe^x analogues or partially glycosidated structures.

Elices: There is some indication that SLe^x might be a ligand for L-selectin as well. Do the results of your neutrophil migration/margination experiments contradict this?

Etzioni: I don't think we can say one way or the other. Still, I believe that induced SLe^x is involved in L-selectin binding.

Hynes: I thought the evidence from Rosen's lab (Imai et al 1993) about the ligand for L-selectin leaned towards the sulphated sialylated sugars rather than SLe^x.

Etzioni: Yes, that's true.

Wagner: Could you look non-invasively at the size of the lymph nodes in the children?

Etzioni: The size of the lymph nodes is normal.

Wagner: This would indicate that homing of lymphocytes is normal.

Labow: Why is there a high turnover rate of the neutrophils? Is there some metabolic defect that may be tied into the presentation of SLe^x that causes the poor survival of the patients?

Etzioni: We think that because of the fucose there is some metabolic defect which causes increased turnover and very short half-life of the neutrophils, but we don't know what it is.

Haskard: With regard to the T cell homing *in vivo*, you made the point that the patients were anergic and that this might be evidence for abnormal T cell homing. Have you actually done biopsies to establish that there aren't any lymphocytes in those lesions? The reduced delayed-hypersensitivity response might be more related to an abnormality of monocyte trafficking.

Etzioni: In two months we will have the answer because we did the experiment last week. I believe that the anergy is due to a deficiency of the cutaneous lymphocyte antigen (CLA), which is important in the delayed-type hypersensitivity response (Picker et al 1993).

Pober: When you define these children as anergic, you are defining it by a very specific read-out, which is skin reaction. This has two consequences. First, anergy typically refers to a T cell defect of antigen recognition; you really don't know that you've got a cognitive T cell defect *per se*. Second, even if you think that the disease produces an effector rather than cognitive defect, the skin may be a particularly compromised organ, because there are indications that the lymphocytes that are involved in inflammatory reactions in the skin depend on E-selectin recognition, whereas lymphocytes involved in inflammatory reactions in other organs may have a much lesser dependence (reviewed in Butcher 1991). In other words, these children may have a skin-specific effector defect.

Etzioni: I agree with you, but this is the only thing we can do.

Barker: One of the difficulties of working on human subjects is getting the appropriate people to study. Have any other individuals been identified with LAD II? It would be nice to get hold of some adults so that you can screen them more clearly.

Etzioni: Not that I know of. Several people have asked me to send them our antibodies because they thought they had LAD II patients, but I haven't heard any more.

Barker: Have blood transfusion laboratories been screening for LAD II?

Etzioni: Bombay blood phenotype is not common, but it does occur. Out of several hundred people with Bombay phenotype, no case resembling LAD II was reported. It is likely that there aren't other LAD II patients because they are so unique in their presentations that they would be reported as a special syndrome even without doing antibody studies. We did a literature search and couldn't find anything that resembled our cases.

Hynes: Are the families of the two patients related in any way? Is this the same mutation?

Etzioni: Both families are Moslem Arabs. They live far apart. We tried to connect the two families and we were able to go about 200 years back in the family trees without finding any connection. On the other hand, if those are really the only two cases in the world, I'm sure that four or five hundred years ago they must have been connected.

Hynes: Do you find other cases showing up in the anecdotal histories of these families?

Etzioni: No.

Sanchez-Madrid: The only patient that we have found with LAD I in Madrid had a single point mutation which, coincidentally, was exactly the same mutation as occurred in an LAD I patient from a Mexican family in Texas. In two completely unrelated families the same nucleotide was mutated. No other mutation that has been found for LAD I affects the same nucleotide.

Hynes: It's very surprising that LAD II hasn't shown up somewhere else, because it sounds like a fairly dramatic phenotype.

Etzioni: One of our patients had a very severe gingivitis, just as is often found in LAD I patients, and when a dentist in Seattle saw the child without knowing anything about him, he immediately thought it was LAD I. We asked many dentists whether they had taken care of kids with severe chronic gingivitis. There was a report from Egypt about some children like this but when they measured their SLe^x it was normal.

Verrando: These patients have skin problems. Did you check for the expression of integrins in their skin? These could be related to the poor wound healing.

Etzioni: We haven't looked at this.

Sanchez-Madrid: Is there any defect in the leukocytes of these patients in mediating the interaction of $\alpha_L\beta_2$ integrins with their ICAM-1 ligand on endothelial cells? Having described that the interaction of neutrophils with E-selectin activates $\alpha_L\beta_2$, what happens with the leukocyte integrins in LAD II patients?

Etzioni: In the assays we have looked at, with 12-*O*-tetradecanoyl phorbol-13-acetate (also known as PMA) and other leukocyte activators, an increase in β_2 integrins was observed.

Sanchez-Madrid: In the adhesion assays, did you find that the stickiness of these leukocytes is reduced?

Etzioni: A small fraction of leukocytes which roll can also stick normally. Transmigration is completely normal under static conditions in LAD II; therefore, I don't think there is a defect in the integrin–ICAM-1 interaction.

References

Butcher EC 1991 Leukocyte–endothelial cell recognition: three (or more) steps to specificity and diversity. Cell 67:1033–1036

Etzioni A, Gershoni-Baruch R, Harlan JM 1994 Sialyl Lewis X and neutrophil aggregation. Blood 83:876–877

Imai Y, Lasky LA, Rosen SD 1993 Sulphation requirement for GlyCAM-1, an endothelial ligand for L-selectin. Nature 361:555–557

Picker LJ, Treer JR, Ferguson-Darnell B, Collins PA, Bergstreeser RR, Terstappen LW 1993 Control of lymphocyte recirculation in man. II. Differential regulation of the cutaneous lymphocyte-associated antigen, a tissue-selective homing receptor for skin-homing T cells. J Immunol 150:1122–1136

Santos-Benito FF, Fernandez-Meyorales A, Martin-Lornas M, Nieto-Sampedro M 1992 Inhibition of proliferation of normal and transformed neural cells by blood group-related oligosaccharides. J Exp Med 176:915–918

Leukocyte and endothelial adhesion molecules in ischaemia/reperfusion injuries

Robert K. Winn, Sam R. Sharar, Nicholas B. Vedder and John M. Harlan

Department of Surgery, University of Washington School of Medicine, Harborview Medical Center, 325 Ninth Avenue, ZA-16, Seattle, WA 98104-2499, USA

Abstract. Tissue ischaemia and/or reperfusion cause some of the injury seen in several clinical disorders and are responsible for considerable mortality and morbidity in humans. Part of the injury occurring after reperfusion of ischaemic tissue is the result of interactions between leukocytes adhering to vascular endothelium. Blocking the function of the leukocyte adhesion β_2 integrin complex (CD11/CD18) leads to improved outcome following ischaemia and reperfusion. Functional blockade of either P-selectin or L-selectin can also lead to improved outcome. β_2 integrin blockade prevents firm leukocyte–endothelial adherence, whereas blocking P-selectin or L-selectin prevents leukocyte rolling. Blocking leukocyte adherence at one of several levels may provide improved outcome in a variety of diseases associated with ischaemia and reperfusion.

1995 Cell adhesion and human disease. Wiley, Chichester (Ciba Foundation Symposium 189) p 63–76

Ischaemia and reperfusion injury is a component of a variety of clinical diseases that are responsible for significant mortality and morbidity throughout the Western world. These clinical disorders include stroke, myocardial infarction, organ transplantation, peripheral vascular disease and resuscitation from haemorrhagic shock. Clearly, a portion of the ischaemia/reperfusion injury can occur during the ischaemic period in these diseases; this can lead to complete failure of an organ if the ischaemic period is long enough. Prevention of cellular death from this cause is achieved by the rapid re-establishment of perfusion to the ischaemic region. However, further injury may occur as a result of the necessary reperfusion, thereby presenting somewhat of a dilemma. The reperfusion portion of the injury must result from the host response to the ischaemic injury and, by virtue of its time course, may be preventable by drug intervention.

Neutrophils were implicated by Metchnikoff (1887) more than a hundred years ago as being capable of causing host injury as they perform their normal physiological function of killing bacteria and cleaning up dead cells. In this theory, inflammatory molecules produced and normally released by neutrophils into intracellular phagosomes are suspected of causing injury as a result of their release extracellularly. Clearly, neutrophils have the capacity to produce proteases and toxic oxygen products that can cause tissue damage, and they have been implicated in the reperfusion portion of ischaemia/reperfusion injuries. *In vitro* experiments have shown that the adherence of neutrophils to the vascular endothelium is necessary if they are to cause injury (reviewed in Winn et al 1993a). That is, the high concentration of circulating anti-inflammatory molecules overwhelms the inflammatory molecules released extracellularly by neutrophils unless these neutrophil-derived mediators are released into a protected microenvironment where their concentration can exceed that of the anti-inflammatory molecules.

Phagocyte adhesion molecules

The mechanisms of leukocyte adhesion to endothelial cells have been widely studied. Phagocyte (neutrophil and monocyte)–endothelial interactions are of particular interest in ischaemia/reperfusion injuries. These studies have led to the identification of specific molecules on phagocytes and endothelial cells that are responsible for their heterotypic adherence. Three families of molecules have been identified and characterized: the integrin family, the selectin family and the immunoglobulin supergene family of adhesion molecules (reviewed by Harlan et al 1992, Lobb 1992, Paulson 1992, Smith 1992). Briefly, there are two subfamilies (β_1 and β_2) within the integrin family that participate in phagocyte–endothelial cell adhesion. These families of adhesion molecules are made up of a common β-chain non-covalently linked to a heavier α-chain. The β_1 family is found on monocytes but not neutrophils and comprises the β_1 integrins (also known as the very late after activation antigens [VLA]-1–6). Each β_1 receptor consists of a distinct α-subunit (α_1–α_6; CD49a–f) linked to a β_1-subunit (CD29) (Albelda & Buck 1990, Carlos et al 1991, Hemler et al 1990). The β_1 integrins are mainly responsible for leukocyte adherence to the extracellular matrix; however, $\alpha_4\beta_1$ (also known as VLA-4 and CD49d/CD29) is responsible for the adherence of mononuclear phagocytes (Carlos et al 1991, Rice et al 1990), lymphocytes (Rice et al 1990, Schwartz et al 1990, Walsh et al 1991), eosinophils (Bochner et al 1991, Dobrina et al 1991, Walsh et al 1991) and basophils (Bochner et al 1991) to endothelial cells. $\alpha_4\beta_1$ has been shown to recognize endothelial vascular cell adhesion molecule 1 (VCAM-1, also known as CD106), a member of the immunoglobulin supergene family of molecules (Elices et al 1990).

The β_2 family of integrins consists of three heterodimers with three separate α-subunits (α_L, CD11a; α_M, CD11b; α_X, CD11c) associated with the β_2-subunit (CD18). These molecules are found exclusively on leukocytes and mediate leukocyte–endothelial cell adherence. $\alpha_L\beta_2$ has been shown to recognize the endothelial ligands intercellular adhesion molecule 1 (ICAM-1, also known as CD54) and ICAM-2 (CD102) (Marlin & Springer 1987, Simmons et al 1988) (members of the immunoglobulin supergene family of adhesion molecules) on endothelial cells and $\alpha_M\beta_2$ recognizes ICAM-1 (Diamond et al 1990, Smith et al 1988). The selectin family consists of three molecules designated P-selectin, L-selectin and E-selectin. L-selectin is found on leukocytes, while P-selectin and E-selectin are both found on endothelial cells. These molecules recognize carbohydrate counter structures including the sialyl Lewis X (SLe^x) antigen (Arnaout 1990, Paulson 1992). L-selectin binds to CD34 on high endothelial venules of peripheral lymph nodes (Baumheter et al 1993) but the counter structure to the phagocyte L-selectin on systemic endothelial cells has not been identified. P-selectin and E-selectin bind to the appropriately sialylated/fucosylated protein structure on phagocytes (Sako et al 1993).

In vivo observations by intravital microscopy have revealed a sequence of events leading to leukocytes arriving at sites of inflammation. The initial step is leukocyte rolling along blood vessel walls, which is followed by firm adherence, then migration along the vessel wall and diapedesis between endothelial cells (Arfors et al 1987). The selectins have been shown to mediate early rolling (Abbassi et al 1993, Lawrence & Springer 1991, Ley et al 1991, von Andrian et al 1991), but they cannot initiate rolling or firm adherence in the face of a sufficiently high shear stress. The β_2 integrin complex is responsible for the firm leukocyte adherence (Arfors et al 1987), but at normal shear stress it cannot support rolling. Thus, the present working model for leukocyte emigration requires two steps, each involving a different family of adhesion molecules (von Andrian et al 1991). Blocking either of these adhesive pathways should therefore be effective in preventing leukocyte emigration.

Ischaemia/reperfusion injury and phagocyte adherence

Experiments in which neutrophils were depleted by various means have established the role of polymorphonuclear leukocytes in tissue injury following ischaemia and reperfusion (reviewed in Winn et al 1993b). Ischaemia/reperfusion injuries of the heart, gut, liver and whole body have been shown to be reduced following neutrophil depletion.

Monoclonal antibodies which block specific adhesion molecules have been the primary tools used for studying the role of leukocyte adhesion in vascular and tissue damage caused by ischaemia/reperfusion. In one of the first experiments of this kind, Hernandez et al (1987) treated cats with anti-β_2 integrin monoclonal antibodies prior to producing ischaemia of the mesentery.

They evaluated the extent of ischaemic injury by measuring the change in permeability following reperfusion. Treatment with the anti-β_2 monoclonal antibodies ameliorated the increased permeability, indicating a significant reduction in microvascular injury. Other early reports described therapy with antibodies raised against adhesion molecules in reperfusion of ischaemic myocardium (Simpson et al 1988) and in haemorrhagic shock (Vedder et al 1988). The latter is thought to represent whole body ischaemia/reperfusion. Myocardial ishcaemia was investigated using anti-α_M monoclonal antibodies, with administration after ischaemia was established but before reperfusion. Those animals treated with monoclonal antibodies 45 min into a 90 min ischaemic period had a reduced area of infarction, measured as a percentage of the area at risk (Simpson et al 1988). Pretreatment with anti-β_2 monoclonal antibodies reduced injury to the stomach and liver as well as reducing mortality as measured 7 days after haemorrhage (Hernandez et al 1987). When haemorrhagic shock was treated at the time of resuscitation, organ injury and mortality were significantly improved in animals treated with anti-β_2 antibodies (Vedder et al 1989).

Reperfusion was shown to be primarily responsible for ischaemia/reperfusion injury through the use of anti-β_2 monoclonal antibodies given either before ischaemia or at the time of reperfusion (Vedder et al 1990). In these experiments, rabbit ears were made ischaemic, then perfusion was re-established and the animals were divided into three treatment groups. One additional group served as a 'sham' control in that they were subjected to the same surgical procedure, but without ischaemia. In the three ischaemic groups, one was treated with saline and the other two with anti-β_2 monoclonal antibodies either prior to ischaemia or at the time of reperfusion. Neither antibody-treated group was statistically different from the sham group and both had significantly less injury than the saline-treated control group. In summary, these early studies established both the role of leukocyte adhesion molecules and the importance of reperfusion in the production of ischaemia/reperfusion injury.

Subsequently, ischaemia/reperfusion injury of other organs and under different conditions has been investigated using reagents that inhibit the adhesive function of α_L, α_M, β_2, P-selectin, L-selectin, ICAM-1 and SLe^x. Some of the *in vivo* ischaemia/reperfusion injuries that have been reported in recent years are listed below. For a review of earlier work see Harlan et al (1992).

Myocardial protection

Selectin inhibition has been accomplished by anti-L-selectin antibodies (Ma et al 1993), anti-P-selectin antibodies (Weyrich et al 1993) and SLe^x-like carbohydrates (Buerke et al 1994), following occlusion of the left anterior descending coronary artery. Treatment was given 10 min prior to release of a 90 min vascular occlusion. Necrosis (as a percentage of the area at risk) was determined at a relatively early time point (after 4 h of reperfusion); this was found to be reduced in treated animals. The protection seen in these experiments is consistent with the theory that leukocyte rolling precedes firm attachment and that rolling is mediated by either L-selectin or P-selectin in ischaemia/reperfusion: blocking either of these molecules is sufficient to prevent significant injury to the heart.

Surgical procedures are frequently performed that require cooling of the heart and arrest of perfusion for extended periods. This results in cold ischaemia; reperfusion results in some degree of myocardial dysfunction. Anti-β_2 monoclonal antibodies given prior to ischaemia were shown to improve myocardial function when compared with the control treatment (Kawata et al 1992). Likewise, the blockade of α_M ameliorated the myocardial injury resulting from low-temperature ischaemia/reperfusion (Wilson et al 1993).

Cerebral injury

Brain injury following embolic stroke was reduced by anti-ICAM-1 monoclonal antibodies given 5 min after initiation of the stroke (Bowes et al 1993). This was just as effective as treatment with tissue plasminogen activator given 30 min after the start of the embolic stroke. The combination of anti-ICAM-1 antibodies and tissue plasminogen activator was no more effective than each treatment alone. It is not clear whether the injury in these experiments is the result of ischaemia or reperfusion. The tissue plasminogen activator should relieve ischaemia but it is not clear that a reperfusion injury occurred in this setting. The anti-ICAM-1 antibody treatment may have prevented leukocyte adherence, thereby preventing ischaemic injury. In a separate study, Okada et al (1994) produced brain ischaemia by the occlusion of cerebral blood vessels, followed by reperfusion 1 h later. They noted persistent expression of P-selectin during ischaemia and increased expression of ICAM-1 starting 1 h after reperfusion. Presumably, these endothelial molecules were responsible for increased adherence of leukocytes under these conditions.

Remote lung injury

Remote injury to the lung has been observed following ischaemia and reperfusion of the rat gut (Hill et al 1992, 1993). Monoclonal antibodies raised against both α_M (Hill et al 1992) and β_2 (Hill et al 1993) were effective in preventing increased permeability of the lung, but did not reduce neutrophil accumulation in that organ following gut reperfusion. Seekamp et al (1993) reported similar distant-organ injury to the lung following reperfusion of ischaemic lower extremities in rats. The increased permeability and haemorrhage in the lung following rat hind-limb ischaemia was reduced by antibodies against α_L, α_M, β_2 and ICAM-1 (Seekamp et al 1993). Each of these monoclonal antibodies significantly reduced leukocyte sequestration in the lungs. The reason for the differences in neutrophil sequestration following gut and hind limb ischaemia is not known.

Haemorrhagic shock

Previous studies in our laboratory have shown that anti-β_2 monoclonal antibodies are effective in preventing organ injury and death following haemorrhagic shock in rabbits and non-human primates (Mileski et al 1990, Vedder et al 1988, 1989). Haemorrhagic shock reduced cardiac output to 30% of baseline for 1 h (Vedder et al 1988) or 2 h (Vedder et al 1989) in rabbits that were resuscitated with their own shed blood and additional lactated Ringer's solution for the next 4 h to maintain their cardiac output at baseline values. The anti-β_2 antibody was given either just before haemorrhage (Vedder et al 1988) or just before resuscitation (Vedder et al 1989). Both treatment protocols resulted in improved survival. Non-human primates had their cardiac output reduced to 30% of baseline for 90 min by haemorrhagic shock and then were resuscitated with their shed blood and lactated Ringer's solution to maintain their cardiac output near baseline (Mileski et al 1990). The 24 h fluid requirements were significantly less than those for the saline-treated control group.

Soft tissue ischaemia

We have examined ischaemia/reperfusion of rabbit ears using monoclonal antibodies raised against β_2 integrin (Vedder et al 1990), P-selectin (Winn et al 1993c) and L-selectin (Mihelcic et al 1994). In these experiments, the rabbit ear was partially amputated at the base leaving only the central artery, vein and a cartilage bridge intact. We then placed a microvascular clamp across the central artery. In the initial experiments, the ischaemic period lasted for 10 h at 22°C ambient temperature (Vedder et al 1990). Tissue oedema was measured daily and tissue necrosis was estimated at the completion of the experiment. Treatment with anti-β_2 monoclonal antibody was either just prior to ischaemia or at the time of reperfusion. Ear oedema in the anti-β_2 antibody-treated groups was similar, and in both was significantly less than in the saline-treated group. We subjected one additional group to the operation but not ischaemia; the extent of oedema in this group was similar to that in two groups treated with anti-β_2 antibody. Tissue necrosis was reduced in the anti-β_2 treated groups compared with the saline group.

In the groups treated with the anti-selectin antibodies, the ischaemic period was 6 h with the ambient temperature adjusted to 23.5–24°C (Mihelcic et al 1994, Winn et al 1993c). Injury was measured in the same manner as in the previous studies. The animals treated with the anti-selectin monoclonal antibodies developed less oedema than animals treated with control monoclonal antibodies that do not block selectin function in rabbits. Likewise, tissue necrosis was decreased in the anti-selectin-treated rabbits compared with the saline-treated group. Myeloperoxidase concentration was measured as an indicator of neutrophil infiltration into the injured ear; this was lower in the antibody-treated groups.

Summary

The experiments described above demonstrate that leukocytes are responsible for a considerable portion of the injury that occurs as a result of reperfusion of ischaemic tissue in a number of different organs. In general, these results are in agreement with the current model of initial leukocyte rolling mediated by selectins and subsequent firm leukocyte adherence as a result of β_2 integrin interaction with ICAM-1. Specifically, blocking of the selectins or inhibition of β_2 or ICAM-1 appears to produce equivalent attenuation of reperfusion injury.

References

Abbassi O, Kishimoto TK, McIntire LV, Anderson DC, Smith CW 1993 E-selectin supports neutrophil rolling *in vitro* under conditions of flow. J Clin Invest 92:2719–2730
Albelda S, Buck C 1990 Integrins and other cell adhesion molecules. FASEB (Fed Am Soc Exp Biol) J 4:2868–2880
Arfors K-E, Lundberg C, Lindbom L, Lundberg K, Beatty PG, Harlan JM 1987 A monoclonal antibody to the membrane glycoprotein complex CD18 inhibits polymorphonuclear leukocyte accumulation and plasma leakage *in vivo*. Blood 69:338–340
Arnaout MA 1990 Structure and function of the leukocyte adhesion molecules CD11/CD18. Blood 75:1037–1050
Baumheter S, Singer MS, Henzel W et al 1993 Binding of L-selectin to the vascular sialomucin CD34. Science 262:436–438
Bochner BS, Luscinskas FW, Gimbrone MA et al 1991 Adhesion of human basophils, eosinophils, and neutrophils to interleukin 1-activated human vascular endothelial cells: contribution of endothelial cell adhesion molecules. J Exp Med 173:1553–1556
Bowes MP, Zivin JA, Rothlein R 1993 Monoclonal antibody to the ICAM-1 adhesion site reduces neurological damage in a rabbit cerebral embolism stroke model. Exp Neurol 119:215–219
Buerke M, Weyrich AS, Zheng Z, Gaeta FCA, Forrest MJ, Lefer AM 1994 Sialyl Lewis X-containing oligosaccharide attentuates myocardial reperfusion injury in cats. J Clin Invest 93:1140–1148
Carlos T, Kovach N, Schwartz B et al 1991 Human monocytes bind to two cytokine-induced adhesive ligands on cultured human endothelial cells: endothelial–leukocyte adhesion molecule-1 and vascular cell adhesion molecule-1. Blood 77:2266–2271
Diamond MS, Staunton DE, de Fougerolles AR et al 1990 ICAM-1 (CD54): a counter-receptor for Mac-1 (CD11b/CD18). J Cell Biol 111:3129–3139
Dobrina A, Menegazzi R, Carlos T et al 1991 Mechanisms of eosinophil adherence to cultured vascular endothelial cells. J Clin Invest 88:20–26

Elices MJ, Osborn L, Takada Y et al 1990 VCAM-1 on activated endothelium interacts with the leukocyte integrin VLA-4 at a site distinct from the VLA-4/fibronectin binding site. Cell 60:577–584

Harlan JM, Winn RK, Doerschuk CM, Vedder NB, Rice CL 1992 *In vivo* models of leukocyte adherence to endothelium. In: Harlan JM, Liu D (eds) Adhesion: its role in inflammatory disease. WH Freeman, New York, p 117–150

Hemler M, Elices M, Parker C, Takada Y 1990 Structure of the integrin VLA-4 and its cell–cell and cell–matrix adhesion function. Immunol Rev 114:45–65

Hernandez LA, Grisham MB, Twohig B, Arfors K-E, Harlan JM, Granger DN 1987 Role of neutrophils in ischemia-reperfusion induced microvascular injury. Am J Physiol 253:H699–H703

Hill J, Lindsay T, Rusche J, Valeri CR, Shepro D, Hechtman HB 1992 A Mac-1 antibody reduces liver and lung injury but not neutrophil sequestration after intestinal ischemia-reperfusion. Surgery 112:166–172

Hill J, Lindsay T, Valeri CR, Shepro D, Hechtman HB 1993 A CD18 antibody prevents lung injury but not hypotension after intestinal ischemia-reperfusion. J Appl Physiol 74:659–664

Kawata H, Aoti M, Hickey PR, Mayer JE Jr 1992 Effect of antibody to leukocyte adhesion molecule CD18 on recovery of neonatal lamb hearts after 2 hours of cold ischemia. Circulation (suppl 5) 86:364–370

Lawrence MB, Springer TA 1991 Leukocytes roll on a selectin at physiologic flow rates: distinction from and prerequisite for adhesion through integrins. Cell 65: 859–873

Ley K, Gaehtgens P, Fennie C, Singer MS, Lasky LA, Rosen SD 1991 Lectin-like cell adhesion molecule 1 mediates leukocyte rolling in mesenteric venules in vivo. Blood 77:2553–2555

Lobb RR 1992 Integrin-immunoglobulin superfamily interactions in endothelial–leukocyte adhesion. In: Harlan JM, Liu DY (eds) Adhesion: its role in inflammatory disease. WH Freeman, New York, p 1–18

Ma XL, Weyrich AS, Lefer DJ et al 1993 Monoclonal antibody to L-selectin attenuates neutrophil accumulation and protects ischemic reperfused cat myocardium. Circulation 88:649–658

Marlin SD, Springer TA 1987 Purified intercellular adhesion molecule-1 (ICAM-1) is a ligand for lymphocyte function-associated antigen 1 (LFA-1). Cell 51:813–819

Metchnikoff E 1887 Sur lallutte des cellules de l'organisme contre l'invasion des microbes. Ann Inst Pasteur 1:321

Mihelcic D, Schleiffenbaum B, Tedder TF, Harlan JM, Winn RK 1994 Inhibition of leukocyte L-selectin function with a monoclonal antibody attenuates reperfusion injury to the rabbit ear. Blood, submitted

Mileski WJ, Winn RK, Vedder NV, Pohlmann TH, Harlan JM, Rice CL 1990 Inhibition of CD18-dependent neutrophil adherence reduces organ injury after hemorrhagic shock in primates. Surgery 108:205–212

Okada Y, Copeland BR, Mori E, Tung MM, Thomas WS, del-Zoppo GJ (1994) P-selectin and intercellular adhesion molecule-1 expression after focal brain ischemia and reperfusion. Stroke 25:202–211

Paulson JC 1992 Selectin/carbohydrate-mediated adhesion of leukocytes. In: Harlan JM, Liu D (eds) Adhesion: its role in inflammatory disease. WH Freeman, New York, p 19–42

Rice GE, Munro JM, Bevilacqua MP 1990 Inducible cell adhesion molecule 110 (INCAM-110) is an endothelial receptor for lymphocytes. A CD11/CD18-independent adhesion mechanism. J Exp Med 171:1369–1374

Sako D, Chang XJ, Barone KM et al 1993 Expression cloning of a functional glycoprotein ligand for P-selectin. Cell 75:1179–1186

Schwartz BR, Wayner EA, Carlos TM, Ochs HD, Harlan JM 1990 Identification of surface proteins mediating adherence of CD11/CD18-deficient lymphoblastoid cells to cultured human endothelium. J Clin Invest 85:2019–2022

Seekamp A, Mulligans MS, Till GO et al 1993 Role of β_2 integrins and ICAM-1 in lung injury following ischemia-reperfusion of rat hind limbs. Am J Pathol 143: 464–472

Simmons D, Makgoba MW, Seed B 1988 ICAM, an adhesion ligand of LFA-1, is homologous to the neural cell adhesion molecule NCAM. Nature 331:624–627

Simpson PJ, Todd RF III, Fantone JC, Mickelson JK, Griffin JD, Lucchesi BR 1988 Reduction of experimental canine myocardial reperfusion injury by a monoclonal antibody (Anti-Mo-1, Anti-CD 11b) that inhibits leukocyte adhesion. J Clin Invest 81:624–629

Smith CW 1992 Transendothelial migration. In: Harlan JM, Liu DY (eds) Adhesion: its role in inflammatory disease. WH Freeman, New York, p 83–115

Smith CW, Rothlein R, Hughes BJ et al 1988 Recognition of an endothelial determinant for CD18-dependent human neutrophil adherence and transendothelial migration. J Clin Invest 82:1746–1756

Vedder NB, Winn RK, Rice CL, Chi E, Arfors K-E, Harlan JM 1988 A monoclonal antibody to the adherence promoting leukocyte glycoprotein CD18 reduces organ injury and improves survival from hemorrhagic shock and resuscitation in rabbits. J Clin Invest 81:939–944

Vedder NB, Fouty BW, Winn RK, Harlan JM, Rice CL 1989 Role of neutrophils in generalized reperfusion injury associated with resuscitation from shock. Surgery 106:509–516

Vedder NB, Winn RK, Rice CL, Chi E, Arfors K-E, Harlan JM 1990 Inhibition of leukocyte adherence by anti-CD18 monoclonal antibody attenuates reperfusion injury in the rabbit ear. Proc Natl Acad Sci USA 81:939–944

von Andrian UH, Chambers JD, McEvoy LM, Bargatze RF, Arfors K-E, Butcher EC 1991 Two-step model of leukocyte–endothelial cell interaction in inflammation: distinct roles for LECAM-1 and the leukocyte β_2 integrins in vivo. Proc Natl Acad Sci USA 88:7538–7542

Walsh GM, Mermod J-J, Hartnell A, Kay AB, Wardlaw AJ 1991 Human eosinophil, but not neutrophil, adherence to IL-1-stimulated human umbilical vascular endothelial cells in $\alpha_4\beta_1$ (very late antigen-4) dependent. J Immunol 146:3219–3423

Weyrich AS, Ma X-L, Albertine KH, Lefer AM 1993 In vivo neutralization of P-selectin protects feline heart and endothelium in myocardial ischaemia and reperfusion injury. J Clin Invest 91:2620–2629

Wilson I, Gillinov AM, Curtis WE et al 1993 Inhibition of neutrophil adherence improves postischaemic ventricular performance of the neonatal heart. Circulation 88: II372–II379

Winn RK, Vedder NB, Mihelcic D, Flaherty LC, Langdale L, Harlan JM 1993a The role of adhesion molecules in reperfusion injury. Agents Action Suppl 41: 113–126

Winn RK, Mihelcic D, Vedder NB, Sharar SR, Harlan JM 1993b Monoclonal antibodies to leukocyte and endothelial adhesion molecules attenuate ischemia-reperfusion injury. Behring Inst Mitt 92:229–237

Winn RK, Liggitt D, Vedder NB, Paulson JC, Harlan JM 1993c Anti-P-selectin monoclonal antibody attenuates reperfusion injury to the rabbit ear. J Clin Invest 92:2042–2047

DISCUSSION

Pals: Your ischaemia/reperfusion results seem very promising for patients who are reperfused, particularly those with acute myocardial infarction. Are any trials underway to test this sort of therapy?

Winn: Not to my knowledge. I know a couple of companies that are working on the anti-β_2 antibody, but these trials have not yet begun. For me, at least, that's a fairly disappointing outcome, but it's not my $100 million!

Etzioni: Buerke et al (1994) have had very promising results using carbohydrates to conserve myocardial tissue in experiments where they have taken myocardial infarction in cats as a model for reperfusion injury.

Hynes: What does it take to get from here to clinical trials?

Winn: I wish I knew! I have heard of several companies that are interested in using carbohydrates to treat ischaemia/reperfusion injuries. However, I don't know of any that are close to clinical trials.

Hynes: But for this sort of acute treatment antibodies would be fine; you wouldn't even have to 'humanize' them.

Winn: That's probably true, although the Food and Drug Administration (FDA) may require antibodies to be humanized. I know that many anti-β_2 antibodies have been humanized; the one that we used has been. The company that owns it dropped the programme. Part of the reason was that they administered the antibody (I don't know whether it was the murine form or the humanized form they used) to two monkeys and they both died of a bleeding problem. I have no explanation for why that occurred.

Elices: Have you looked at the neutrophil counts of animals that you've treated with the anti-β_2 antibodies to see whether there are any differences? The rationale for this is that antibodies tend to hang around for quite some time.

Winn: Bill Mileski did a series of monkey experiments with haemorrhagic shock because we were worried about species specificity (Mileski et al 1990). We followed them up for 48 h, and all of the β_2-treated animals had leukocytosis and neutrophilia at 48 h, much as the LAD I patients; but with the anti-P-selectin treatment we've not followed them that far.

Elices: Would you consider that to be a potential problem for treatment of some of these acute conditions with antibodies?

Winn: All the evidence we have with the ischaemia/reperfusion injury is that the antibodies block a portion of the injury.

Stanley: Perhaps I don't understand the clinical situations well enough, but if I think about what you said for reperfusion injuries, anti-β_2 helps, but it also will increase the chance of infection. Don't the type of injuries we are talking about also have the risk of infection, so that by this treatment you're increasing one risk but decreasing another? Does this have anything to do with why companies may not want to use them for human trials?

Winn: No, I don't think so. With the companies that we've talked to about clinical trials, we've specifically been focusing on trauma patients who are at risk of infection. However, that has not been something that has driven them away from looking at that as an indication, since there is no treatment available for this patient population. Another indication is myocardial infarction: one of the major problems with this population is that you have to have a study group of about 10 000 patients, so you can go broke very quickly—alternatively, you can also get very rich!

Hynes: These patients are really sick, so the risk of infection is probably worth taking.

Winn: That's right. With the trauma patients, we can identify a group that has at least 50% mortality, and there is absolutely no treatment for those patients once they are in hospital. We think their late organ failure is related to the period of hypotension. Our study of more than 150 hypotensive trauma patients who died after they arrived at the hospital included more than 90% of all trauma deaths, with the one criterion of systolic blood pressure less than 90 Torr.

Another approach to therapy of ischaemia/reperfusion injury is to block ICAM-1. This has been shown to be effective in a number of studies. Bill Milesky examined risk of infection and showed that rabbits given anti-ICAM-1 antibody were not at increased risk of infection.

Pals: A study has recently been done in The Netherlands with antibodies against GPIIb/IIIa to treat patients with refractory unstable angina. It appears that anti-GPIIb/IIIa can reduce myocardial infarction and facilitate percutaneous transluminal coronary angioplasty in these patients (Simoons et al 1994).

Winn: I think the anti-GPIIb/IIIa antibody has been approved for use in the USA.

Ruggeri: A study has just been published (Evaluation of 7E3 for the Prevention of Ischaemic Complications [EPIC] 1994). If you read the paper carefully, two things come out pretty clearly. One is that the antibody did what it was supposed to do, which is pleasing; the problem is that it also caused increased bleeding in the patients (which is also what you would have expected). The issue of whether the risk:benefit ratio is such that you would want to use the antibody cannot be definitively addressed by a study of the size of the one published. You need so many patients to do the definitive study that it's questionable whether anybody will ever do it.

Etzioni: From the clinical point of view, it is very important to know the kinetics of the antibodies, because if you treat chronic diseases, you will have to administer the antibodies several times. But here we're speaking about an acute condition in which you have to give antibodies just once.

Winn: I think you have some kind of time window for treatment, but I don't know how long it is. The antibodies will probably be humanized before use; if they are, their half-life may be increased from 12–24 h up to 15 days or so. So you may have a very long half-life with an increased risk of infection.

Pober: How big a study you have to do is dependent on how small an effect you're hoping to see. Part of the problem is that not only is there the biological variability, but also the neutrophil-mediated component of ischaemia/reperfusion is only one part of the injury that these patients and their organs are suffering. First, there are ischaemic (i.e. hypoxic) changes that directly injure parenchymal cells and there are ischaemic changes in endothelial barrier function (i.e. ischaemia tends to open up endothelial junctions)—these alterations will be independent from reperfusion and the neutrophil-mediated component. Second, although it's not clear how relevant it is in humans at this point, there are oxidative injuries that can be produced upon reperfusion that are also independent of neutrophils, mediated through the generation of oxy radicals by the xanthine oxidase system in endothelial cells. This is partly why it's difficult to justify a trial.

Hynes: The trials for this sort of trauma would not be as nearly as large as the ones that Zaverio Ruggeri was talking about.

Rothlein: Presumably, the animal models address those issues too. At least from the industry point of view, the clinicians say that mimicking cardiac reperfusion injury is going to be a very cumbersome trial because it is not sufficient to show that you have spared heart tissue, it is necessary to show a clinical benefit to the patient.

Ruggeri: It's not that the regulatory agencies just don't like admitting facts that physicians consider obvious; the problem is that we may have to find ways to think about these trials in a different light with regard to endpoints. At present the outcome of any trial seems to be the need for another trial: at least in the field of new antithrombotic drugs, this is what is happening.

Rothlein: Which is why industry is reluctant to get into this.

Ruggeri: There seems to be no indication that any of these trials is going to lead to a conclusive, positive result. Nevertheless, the need for improved therapeutic approaches in the field of acute coronary artery disease is obvious, since mortality rates are still pretty high, particularly in patients with complications.

Ruggeri: The problem is that there's no evidence that anyone will be able to change the FDA position, at least in the short term.

Pals: Have you ever used combinations of antibodies to look for additive beneficial effects?

Winn: No, we haven't. In some of the earlier experiments we did a sham operation on the rabbit ear where we simply transected the ear, reattached it, but did not make the ear ischaemic (Vedder et al 1990). We looked at the ear volume in those experiments and it was not different from the anti-β_2-treated rabbits. Most of the injury that we have shown is reperfusion injury. There is some ischaemic injury, but this is probably related to the lack of blood flow to the edges of the ear where we have cut major veins.

Pals: You mentioned that the temperature is very important in ischaemia/reperfusion: of course, if you want to treat a patient it will be at 37 °C and not 24 °C, so why did you use 24 °C?

Winn: 24 °C is the ambient temperature. I can't bring ambient temperature to 37 °C for 6 h in my experiments. If you look at ischaemic skeletal muscle, it can withstand ischaemia for about 1–2 h; again, the degree of injury is temperature dependent. But your point is well taken that 37 °C is the temperature at which most patients would be treated.

Rothlein: From the therapeutic point of view, the mechanism of action of the anti-β_2 antibody is multiple in terms of the reperfusion injury. You could be blocking the aggregation of neutrophils, attachment of neutrophils to endothelial cells, or trafficking of neutrophils. Presumably, the anti-P-selectin would be blocking trafficking of neutrophils. Have you done the kinetic study with anti-P-selectin as you have done with anti-β_2 where you showed therapeutic benefit with treatment delayed up to 4 h after reperfusion?

Winn: No. We've done some preliminary experiments with carbohydrates where we've delayed treatment by up to 4 h. In these experiments there was no difference between the saline control treatment and carbohydrate treatment 4 h after reperfusion. However, treatment 1 h post reperfusion was just as effective as treatment at the time of reperfusion. Treatment with carbohydrate was as effective as with anti-P-selectin. This suggests that if you delay anti-P-selectin treatment by an hour, it should still be very effective.

Rothlein: But it's not as effective as anti-β_2?

Winn: No; at least, the carbohydrates don't seem to be as effective at 2 h and anti-P-selectin also may not be as effective.

Shaltiel: Is it possible to provide the same protection using monovalent Fab fragments or $F(ab)'_2$ fragments of the monoclonal antibody?

Winn: The monovalent fragments are cleared rapidly from the circulation with a half-life of about 30 min, so an infusion of Fab over several hours is needed to ensure saturation of the adhesion molecule. $F(ab)'_2$ fragments are probably as effective as the whole antibody, but we've not done those experiments.

Wagner: I'm intrigued by your published observation that in the presence of anti-P-selectin antibody, injected *Staphylococcus aureus* made a smaller lesion (Sharar et al 1993). Perhaps *S. aureus* uses P-selectin as a vehicle for entry.

Winn: I don't know, it might do.

Etzioni: What was the therapeutic dose in humans? In your rabbit experiments you used a huge dose of antibodies.

Winn: To work out an effective therapeutic dose, you need to know the concentration necessary for saturating the molecule and for how long you need to protect. We used 2 mg/kg because it was the dose that could block neutrophil immigration induced by lipopolysaccharide in rabbits. I don't know if it would still work if you reduced the concentration.

Hynes: From some of your results on the ICAM blockers, it seems that if you knock out part of the system, such as one of the selectins or one of the ligands for β_2 integrins, you don't have so much of a problem with infection and you can proceed with the blocking, whereas when you knock out all three β_2 integrins interacting with any of their ligands at the dose you're using, you're getting problems with infection. It might be better to try a lower dose and have less effective inhibition. Presumably, there is a window where you can inhibit enough neutrophil function, but maybe not all of it.

Winn: Yes, I think you could reduce the dose and still see some beneficial effects.

Hynes: I don't think the dose is out of line; I think companies can make enough antibody for that.

Rothlein: They can, but that means they would have to titre each patient, which would be an impossible situation.

Hynes: Why would they have to titre every patient?

Rothlein: If you're worried about getting a bimodal curve—that if you give too much antibody, the patients are at risk of getting an infection, and if you give too little antibody, there will be no therapeutic benefit—then somehow you have to know exactly how much to give each patient. Because the amount of antibody necessary to achieve the same level in each patient is dependent on variables such as weight, integrin expression and blood counts, then each patient must be individually titred.

I would be interested if somebody did an infection model with a functional anti-$\alpha_M\beta_2$ antibody. I think anti-β_2 is blocking the $\alpha_M\beta_2$ activity on polymorphonuclear leukocytes and that's why the rabbits are prone to infections. I don't think this susceptibility has to do with anti-β_2 inhibition of leukocyte trafficking. My prediction would be that anti-$\alpha_M\beta_2$ would promote infection, anti-β_2 would, and anti-$\alpha_L\beta_2$ wouldn't.

References

Buerke M, Weyrich AS, Zdeng Z, Gaeta FCA, Forrest MJ, Lefer AM 1994 Sialyl-Lewis^x-containing oligosaccharide attenuates myocardial reperfusion injury in cats. J Clin Invest 93:1140-1148

EPIC 1994 Use of a monoclonal antibody directed against the platelet glycoprotein IIb/IIIa receptor in high-risk coronary angioplasty. N Engl J Med 330:956–961

Mileski WJ, Winn RK, Vedder NV, Pohlman TH, Harlan JM, Rice CL 1990 Inhibition of CD18-dependent neutrophil adherence reduces organ injury after hemorrhagic shock in primates. Surgery 108:205–212

Sharar SR, Sasaki SS, Flaherty LC, Paulson JC, Harlan JM, Winn RK 1993 P-selectin blockade does not impair leukocyte host defense against bacterial peritonitis and soft tissue infection in rabbits. J Immunol 151:4982–4988

Simoons ML, DeBeer MJ, Vandenbrand MJBM et al 1994 Randomized trial of GPIIb/IIIa platelet receptor blocker in refractory unstable angina. Circulation 89:596–603

Vedder NB, Winn RK, Rice CL, Chi E, Arfors K-E, Harlan JM 1990 Inhibition of leukocyte adherence by anti-CD18 monoclonal antibody attenuates reperfusion injury in the rabbit ear. Proc Natl Acad Sci USA 81:939–944

General discussion I

Hynes: One thing that occurred to me in listening to this last discussion, comparing the animal models that we heard about first with the more clinical things that we heard about later, is that there is clearly a need for filling in the gaps between the two. There are human diseases for which we still lack good animal models, although I expect these will appear soon. For instance, it would be nice to have an $\alpha_M\beta_2$ knockout, a β_2 knockout and an α_L knockout. I throw this out as a suggestion to those of you who want to knock things out!

Labow: The results from many animal models of adhesion defects are mixed as far as which genes are most important. Very likely it is just simply a matter of the relative levels of expression of the various adhesion molecules in different animals. Picking clinical modalities will probably be easier if, for the various human diseases (trauma, myocardial infarction, strokes, etc.), people catalogue the different adhesion molecules that are expressed in primates. There really are very few data on which gene products are presented in each of these scenarios. This information would help us decide which molecules to target therapeutically. Although β_2 is a great adhesion molecule because it is probably a central player in many of the trafficking reactions, for us to produce treatments that are more specific and less toxic we will need to fine-tune the molecules we target.

Ruggeri: I think this is a very important point; similar issues are faced by people interested in antithrombotic therapy. There, the problem is whether you can ever achieve antithrombotic efficacy without paying the price of bleeding complications. As far as platelets are concerned, the choice of inhibiting GPIIb/IIIa is the one that is most likely to result in bleeding complications, because this is the only platelet receptor that is absolutely required to mediate platelet–platelet interactions under all flow conditions. A more promising approach could be that of targeting pathays that are predominantly involved in platelet–surface interactions under high shear stress conditions; this could lead to significant antithrombotic activity, since (at least in coronaries) thrombi form in a high shear stress environment, with lesser bleeding complications. Obviously, it is difficult to select the appropriate targets for therapeutic inhibition of complex pathophysiological mechanisms.

Elices: I think that kind of thinking is more applicable to antibody therapy. One needs to evaluate the risk:benefit ratio. The animal models that have been talked about here indicate that some of these adhesion molecules are very important, and if you disrupt them using antibodies you are likely to cause other complications. For instance, as you pointed out, antithrombotic therapy with anti-GPIIb/IIIa antibodies causes bleeding. A GPIIb/IIIa inhibitor that is cleared more rapidly than the antibody might not have the same complications.

Ruggeri: This may not be true, because you can't choose the time when you need haemostasis to save you from intracranial bleeding. The study with the anti-GPIIa/IIIb antibody clearly showed that bolus infusion needed to be followed by prolonged constant infusion in order to achieve the best antithrombotic results. Another key player in haemostasis, thrombin, has been targeted because of the appealing idea that if you knock out a crucial agonist you have a better therapeutic effect. Two trials of antithrombins have been stopped because of intracranial bleeding. So, although the idea of targeting key molecules may be very appealing, the concept of being a little bit more selective has the potential for fewer side effects.

Pober: Returning to inflammation and particularly the issue of reperfusion injury, P-selectin may be the molecule of choice. Several years ago, in a study with Heinz Redl and Günther Schlag from Vienna, we compared expression of E-selectin in polytrauma/haemorrhagic shock with septic shock (Redl et al 1991). Whereas septic shock produced massive up-regulation of E-selectin, there was very little E-selectin to be found in animals that were subjected to polytrauma/haemorrhagic shock. These data suggest that traumatic shock is a case where P-selectin is probably a better target than E-selectin. Since E-selectin is not inhibited, you may avoid the complication of subsequent sepsis where there may be more functional redundancy between the two molecules.

Hynes: Which comes back to the Mark Labow's point that we really need to know which adhesion molecules are expressed where in all these situations.

Reference

Redl H, Dinges HP, Buurman WA et al 1991 Expression of endothelial leukocyte adhesion molecule-1 (ELAM-1) in septic but not traumatic/hypovolemic shock in the baboon. Am J Pathol 139:461–466

The integrin $\alpha_4\beta_1$ (VLA-4) as a therapeutic target

Mariano J. Elices

Cytel Corporation, 3525 John Hopkins Court, San Diego, CA 92121, USA

Abstract. Disease models in animals demonstrate that the leukocyte integrin $\alpha_4\beta_1$ (VLA-4) is a suitable target for therapy in a number of chronic inflammatory disorders. While *in vivo* studies have concentrated on the use of anti-α_4 antibodies as proof of concept tools, repeated administration to combat human chronic inflammatory conditions is likely to require small antagonists of $\alpha_4\beta_1$. We have developed low molecular weight $\alpha_4\beta_1$ inhibitors which have shown therapeutic promise in animal models of chronic inflammation.

1995 Cell adhesion and human disease. Wiley, Chichester (Ciba Foundation Symposium 189) p 79–90

Leukocyte and endothelial cell adhesion molecules have increasingly become attractive targets for therapeutic intervention in inflammation. This stems from the act that during the inflammatory cascade, adhesion molecules play a key role in both recruitment of circulating white blood cells and their subsequent extravasation into inflamed tissues. Thus, cell adhesion blockade constitutes an early therapeutic target to interfere with the untoward effects of inflammation.

Integrins, cell adhesion molecules known to participate in the inflammatory cascade, are generally regarded as a family of versatile and widely distributed cell adhesion receptors (Hynes 1992). We have focused on the leukocyte integrin $\alpha_4\beta_1$ (also known as VLA-4 and CD49d/CD29) as a therapeutic target because it is predominantly expressed on lymphocytes, monocytes and eosinophils (Hemler 1990), the leukocyte populations which primarily mediate chronic inflammation and allergy.

In vitro studies of $\alpha_4\beta_1$ function

At the molecular level, most adhesion functions mediated by $\alpha_4\beta_1$ can be explained by a direct interaction between the $\alpha_4\beta_1$ integrin and either of two separate counter-receptor structures, namely, VCAM-1 (Elices et al 1990, Rice et al 1990, Schwartz et al 1990) and alternatively spliced variants of fibronectin containing CS-1 (Wayner et al 1989, Guan & Hynes 1990). VCAM-1 is a member of the immunoglobulin gene superfamily expressed on a variety of cell types, including vascular endothelium, in response to pro-inflammatory cytokines (Osborn 1990, Gearing & Newman 1993). CS-1 is a 25 amino acid sequence (Humphries et al 1986), which arises by alternative splicing within the IIICS or V region of fibronectin (for a review, see Hynes 1990).

While an $\alpha_4\beta_1$ peptide recognition motif remains to be defined for VCAM-1, a sequence within CS-1 which contains the binding site for $\alpha_4\beta_1$ has been identified (Wayner et al 1989, Guan & Hynes 1990). Truncation analysis has further revealed that the minimum essential sequence for activation-dependent $\alpha_4\beta_1$ recognition of CS-1 is the tripeptide LDV (Komoriya et al 1991, Wayner & Kovach 1992).

Despite the enormous attention bestowed upon VCAM-1, fibronectin variants expressing CS-1 may be key components in inflammation because alternative splicing of CS-1 represents an efficient mechanism for generating $\alpha_4\beta_1$ binding sites in fibronectin. Previously, the process of alternative splicing in fibronectin was shown to be regulated in a cell type-specific manner, especially during cell migration and development (for a review, see Hynes 1990). More recently, CS-1-containing fibronectin has been shown to be selectively expressed on blood vessels in synovial membrane tissue from rheumatoid arthritis patients (Elices et al 1994).

Transendothelial migration of $\alpha_4\beta_1$-expressing leukocytes may be a potential role for variants of fibronectin containing CS-1 in inflammation. For instance, transmigration of eosinophils across endothelial cell monolayers *in vitro* has been reported to be dependent on $\alpha_4\beta_1$ and CS-1 fibronectin (Kuijpers et al 1993). This observation is hardly surprising, since human endothelial cells in culture are known to express alternatively spliced CS-1 variants of fibronectin (Hershberger & Culp 1990, Kocher et al 1990, Kuijpers et al 1993, Elices et al 1994) and up-regulation of these fibronectin variants may occur *in vitro* in cytokine-stimulated vascular endothelium (Elices et al 1994, Magnuson et al 1991, Wang et al 1991). In other species, lymphocyte migration across high endothelial cells appears to involve similar requirements for alternatively spliced CS-1 fibronectin (Ager & Humphries 1990, May & Ager 1992, Hourihan et al 1993).

In vivo studies of $\alpha_4\beta_1$ function

The involvement of the leukocyte integrin $\alpha_4\beta_1$ in chronic inflammatory disorders has been amply demonstrated by studies *in vivo* (see Table 1). These can be arbitrarily classified in two groups: (i) general models of inflammation; and (ii) models of disease. The former provide information on the general role of the integrin $\alpha_4\beta_1$ in inflammation *in vivo*, and constitute the basis for intrinsic-activity, high-throughput *in vivo* assays. A good example is the contact

TABLE 1 Animal models of disease mediated by $\alpha_4\beta_1$

Disease model	*Species*	*Inhibited by*	*References*
Contact hypersensitivity/ delayed-type hypersensitivity	Mouse	GRGDS and GPEILDVPST	Ferguson et al 1991
	Rat	Antibody to α_4	Issekutz 1991a,b, 1993
	Mouse	Antibody to α_4	Elices et al 1993, Chisholm et al 1993, Ferguson & Kupper 1993
	Monkey	Antibody to VCAM-1	Silber et al 1994
Experimental autoimmune encephalomyelitis	Mouse	Antibody to α_4	Yednock et al 1992
	Rat	Antibody to α_4	Baron et al 1993
Nephrotoxic nephritis	Rat	Antibody to α_4	Mulligan et al 1993c, Molina et al 1994
Cutaneous anaphylaxis	Guinea pig	Antibody to α_4	Weg et al 1993
Immune complex-induced lung injury	Rat	Antibody to α_4	Mulligan et al 1993a,b
Spontaneous colitis	Monkey	Antibody to α_4	Podolsky et al 1993
Asthma	Rabbit	Antibody to α_4	Metzger et al 1994
	Rabbit	CS-1 peptide	Metzger et al 1994
	Sheep	Antibody to α_4	Abraham et al 1994
	Rat	Antibody to α_4	Rabb et al 1994
Adjuvant-induced arthritis	Rat	Antibody to α_4	Barbadillo et al 1993
Diabetes	Mouse	Antibody to α_4	Burkly et al 1994

hypersensitivity/delayed-type hypersensitivity *in vivo* model elicited by treatment of the skin with various chemicals (Ferguson et al 1991, Issekutz 1991a,b, 1993, Elices et al 1993, Chisholm et al 1993, Ferguson & Kupper 1993, Silber et al 1994). The immune complex-induced lung injury models also belong to this group (Mulligan et al 1993a,b). The conclusion from these studies is that $\alpha_4\beta_1$ expressed on T lymphocytes, monocytes and eosinophils appears to be involved in recruitment and infiltration, especially in inflammation.

Disease models are more informative about the role of the $\alpha_4\beta_1$ leukocyte integrin under pathological conditions, since these models attempt to mimic as closely as possible features of human disease. Essentially, the skin, brain, kidney, lung and gut are targets of a wide variety of $\alpha_4\beta_1$-dependent inflammatory disorders, mostly resulting from recruitment of mononuclear leukocytes and eosinophils. Thus, a tentative conclusion from these *in vivo* studies is that $\alpha_4\beta_1$ may be involved in multiple sclerosis (Yednock et al 1992, Baron et al 1993), inflammatory bowel disease (Podolsky et al 1993), asthma (Metzger et al 1994, Abraham et al 1994), rheumatoid arthritis (Barbadillo et al 1993) and diabetes (Burkly et al 1994) in humans.

By and large, most of the *in vivo* studies performed to date have employed monoclonal antibodies to thc α_4-subunit of $\alpha_4\beta_1$ (see Table 1). While these have been very useful tools with which to decipher the role of $\alpha_4\beta_1$ *in vivo*, they are unlikely to be practical as therapeutic drugs in some of the potential applications outlined above. This is a major consideration, especially when patients necessitate chronic administration as part of their disease management.

Future objectives

The area of *in vivo* function of $\alpha_4\beta_1$ presents a number of future challenges which I would divide into two general categories, although these goals are not mutually exclusive: (a) basic science; and (b) drug development. In the scientific domain, one would like to understand more precisely the relative roles of the $\alpha_4\beta_1$/VCAM-1 and $\alpha_4\beta_1$/CS-1 adhesion pathways in migration, recirculation and inflammation. In other words, one would like to answer the question of which $\alpha_4\beta_1$ counter-receptor structure, VCAM-1 or CS-1 fibronectin, is more critical in a specific disease. This would also contribute very strongly to the second goal of developing therapies for chronic inflammation in humans, because one could theoretically design specific VCAM-1 or CS-1 antagonists depending on the inflammatory condition to be treated. Towards this latter goal, a CS-1 peptide inhibitor has been shown to be efficacious in an animal model of asthma (Metzger et al 1994) and a cyclic peptide inhibitor of $\alpha_4\beta_1$ has recently been described with the potential for therapeutic use (Nowlin et al 1993). Ultimately, the success of targeting $\alpha_4\beta_1$ for therapeutic applications will be predicated on the drugs that will be developed for human intervention in chronic inflammation.

Acknowledgments

I would like to thank my colleagues Tom Arrhenius, John Fikes, Gary Firestein, Federico Gaeta, Amita Goel, John Harlan, Ya-Bo He, Taco Kuijpers, Jia Lei, Jim Metzger, Jim Paulson, Victoria Ridger, Dana Strahl, Susan Tamraz, Vanessa Tollefson, Van Tsai and Leanne Vollger for their invaluable contributions to this work.

References

Abraham WM, Sielczak MW, Ahmed A et al 1994 α4-integrins mediate antigen-induced late bronchial responses and prolonged airway hyperresponsiveness in sheep. J Clin Invest 93:776–787

Ager A, Humphries MJ 1990 Use of synthetic peptides to probe lymphocyte–high endothelial cell interactions. Lymphocytes recognize a ligand on the endothelial surface which contains the CS1 adhesion motif. Int Immunol 2:921–928

Barbadillo C, Andreu JL, Mulero J, Sanchez-Madrid F 1993 Anti-VLA-4 mAb prevents adjuvant arthritis in Lewis rats. Arthr Rheum (suppl) 36:95 (abstr)

Baron JL, Madri JA, Ruddle NH, Hashim G, Janeway CA 1993 Surface expression of α4 integrin by CD4 T cells is required for their entry into brain parenchyma. J Exp Med 177:57–68

Burkly LC, Jakubowski A, Hattori M 1994 Protection against adoptive transfer of autoimmune diabetes mediated through very late antigen-4 (VLA-4) integrin. Diabetes 43:529–534

Chisholm PL, Williams CA, Lobb RR 1993 Monoclonal antibodies to the integrin α4 subunit inhibit the murine contact hypersensitivity response. Eur J Immunol 23:682–688

Elices MJ, Osborn L, Takada Y et al 1990 VCAM-1 on activated endothelium interacts with the leukocyte integrin VLA-4 at a site distinct from the VLA-4/fibronectin binding site. Cell 60:577–584

Elices MJ, Tamraz S, Tollefson V, Vollger LW 1993 The integrin VLA-4 mediates leukocyte recruitment to skin inflammatory sites in vivo. Clin Exp Rheumatol 11: S77–S80

Elices MJ, Tsai V, Strahl D et al 1994 Expression and functional significance of alternatively spliced CS1 fibronectin in rheumatoid arthritis microvasculature. J Clin Invest 93:405–416

Ferguson TA, Kupper TS 1993 Antigen-independent processes in antigen-specific immunity: a role for α4 integrin. J Immunol 150:1172–1182

Ferguson TA, Mizutani H, Kupper TS 1991 Two integrin-binding peptides abrogate T cell-mediated immune responses in vivo. Proc Natl Acad Sci USA 88:8072–8076

Gearing AJH, Newman W 1993 Circulating adhesion molecules in disease. Immunol Today 14:506–512

Guan J-L, Hynes RO 1990 Lymphoid cells recognize an alternatively spliced segment of fibronectin via the integrin receptor $\alpha_4\beta_1$. Cell 60:53–61

Hemler ME 1990 VLA proteins in the integrin family: structures, functions, and their role on leukocytes. Annu Rev Immunol 8:365–400

Hershberger RP, Culp LA 1990 Cell-type-specific expression of alternatively spliced human fibronectin IIICS mRNAs. Mol Cell Biol 10:662–671

Hourihan H, Allen TD, Ager A 1993 Lymphocyte migration across high endothelium is associated with increases in $\alpha_4\beta_1$ integrin (VLA-4) affinity. J Cell Sci 104:1049–1059

Humphries MJ, Akiyama SK, Komoriya A, Olden K, Yamada KM 1986 Identification of an alternatively spliced site in human plasma fibronectin that mediates cell type-specific adhesion. J Cell Biol 103:2637–2647

Hynes RO 1990 Fibronectins. Springer-Verlag, New York
Hynes RO 1992 Integrins: versatility, modulation, and signalling in cell adhesion. Cell 69:11–25
Issekutz TB 1991a Effect of antigen challenge on lymph node lymphocyte adhesion to vascular endothelial cells and the role of VLA-4 in the rat. Cell Immunol 138:300–312
Issekutz TB 1991b Inhibition of in vivo lymphocyte migration to inflammation and homing to lymphoid tissues by the TA-2 monoclonal antibody: a likely role for VLA-4 in vivo. J Immunol 147:4178–4184
Issekutz TB 1993 Dual inhibition of VLA-4 and LFA-1 maximally inhibits cutaneous delayed-type hypersensitivity-induced inflammation. Am J Pathol 143:1286–1293
Kocher O, Kennedy SP, Madri JA 1990 Alternative splicing of endothelial cell fibronectin mRNA in the IIICS region: functional significance. Am J Pathol 137:1509–1524
Komoriya A, Green LJ, Mervic M, Yamada SS, Yamada KM, Humphries MJ 1991 The minimal essential sequence for a major cell type-specific adhesion site (CS1) within the alternatively spliced type III connecting segment domain of fibronectin is leucine-aspartic acid-valine. J Biol Chem 266:15075–15079
Kuijpers TW, Mul EPJ, Blom M et al 1993 Freezing adhesion molecules in a state of high-avidity binding blocks eosinophil migration. J Exp Med 178:279–284
Magnuson VL, Young M, Schattenberg DG et al 1991 The alternative splicing of fibronectin pre-mRNA is altered during aging and in response to growth factors. J Biol Chem 266:14654–14662
May MJ, Ager A 1992 ICAM-1-independent lymphocyte transmigration across high endothelium: differential upregulation by interferon γ, tumor necrosis factor α, and interleukin 1β. Eur J Immunol 22:219–226
Metzger WJ, Ridger V, Tollefson V, Arrhenius T, Gaeta FCA, Elices M 1994 Anti-VLA-4 antibody and CS-1 peptide inhibitor modifies airway inflammation and bronchial airway hyperresponsiveness (BHR) in the allergic rabbit. J Allergy Clin Immunol 93:125 (abstr)
Molina A, Sanchez-Madrid F, Bricio T et al 1994 Prevention of mercuric chloride-induced nephritis in the brown Norway rat by treatment with antibodies against the α_4 integrin. J Immunol 153:2313–2320
Mulligan MS, Smith CW, Anderson DC et al 1993a Role of leukocyte adhesion molecules in complement-induced lung injury. J Immunol. 150:2401–2406
Mulligan MS, Wilson GP, Todd RF III et al 1993b Role of $\beta1$, $\beta2$ integrins and ICAM-1 in lung injury after deposition of IgG and IgA immune complexes. J Immunol 150:2407–2417
Mulligan MS, Johnson K, Todd RF III et al 1993c Requirements for leukocyte adhesion molecules in nephrotoxic nephritis. J Clin Invest 91:577–587
Nowlin DM, Gorcsan F, Moscinski M, Chiang S-L, Lobl TJ, Cardarelli PM 1993 A novel cyclic pentapeptide inhibits $\alpha_4\beta_1$ and $\alpha_5\beta_1$ integrin-mediated cell adhesion. J Biol Chem 268:20352–20359
Osborn L 1990 Leukocyte adhesion to endothelium in inflammation. Cell 62:3–6
Podolsky DK, Lobb R, King N et al 1993 Attenuation of colitis in the cotton-top tamarin by anti-$\alpha4$ integrin monoclonal antibody. J Clin Invest 92:372–380
Rabb HA, Olivenstein R, Issekutz TB, Renzi PM, Martin JG 1994 The role of the leukocyte adhesion molecules VLA-4, LFA-1, and Mac-1 in allergic airway responses in the rat. Am J Respir Crit Care Med 149:1186–1191
Rice GE, Munro JM, Bevilacqua MP 1990 Inducible cell adhesion molecule 110 (INCAM-110) is an endothelial receptor for lymphocytes. A CD11/CD18-independent adhesion mechanism. J Exp Med 171:1369–1374
Schwartz BR, Wayner EA, Carlos TM, Ochs HD, Harlan JM 1990 Identification of surface proteins mediating adherence of CD11/CD18-deficient lymphoblastoid cells to cultured human endothelium. J Clin Invest 85:2019–2022

Silber A, Newman W, Sasseville VG et al 1994 Recruitment of lymphocytes during cutaneous delayed hypersensitivity in nonhuman primates is dependent on E-selectin and vascular cell adhesion molecule 1. J Clin Invest 93:1554–1563
Wang A, Cohen DS, Palmer E, Sheppard D 1991 Polarized regulation of fibronectin secretion and alternative splicing by transforming growth factor β. J Biol Chem 266:15598–15601
Wayner EA, Kovach NL 1992 Activation-dependent recognition by hematopoietic cells of the LDV sequence in the V region of fibronectin. J Cell Biol 116:489–497
Wayner EA, Garcia-Pardo A, Humphries MJ, McDonald JA, Carter WG 1989 Identification and characterization of the T lymphocyte adhesion receptor for an alternative cell attachment domain (CS-1) in plasma fibronectin. J Cell Biol 109:1321–1330
Weg VB, Williams TJ, Lobb RR, Nourshargh S 1993 A monoclonal antibody recognizing very late activation antigen-4 inhibits eosinophil accumulation *in vivo*. J Exp Med 177:561–566
Yednock TA, Cannon C, Fritz LC, Sanchez-Madrid F, Steinman L, Karin N 1992 Prevention of experimental autoimmune encephalomyelitis by antibodies against $\alpha_4\beta_1$ integrin. Nature 356:63–66

DISCUSSION

Birchmeier: Is the inhibitor of $\alpha_4\beta_1$ you used (Metzger et al 1994) a peptide, or have you modified it?

Elices: It has a peptidic nature, but it is no longer a peptide as it contains organic moieties as well. It's based on the sequence that Richard Hynes and Liz Wayner showed to be involved in $\alpha_4\beta_1$ binding (Wayner et al 1989, Guan & Hynes 1990), but it is modified to contain groups that, for example, permit a longer half-life and confer serum stability.

Sanchez-Madrid: Does your inhibitor of $\alpha_4\beta_1$ also inhibit the interaction between $\alpha_4\beta_7$ and VCAM-1 or CS-1?

Elices: We haven't looked at the $\alpha_4\beta_7$–CS-1 interaction, but I would assume that it is inhibited. From what I understand, the binding event is very similar to that with $\alpha_4\beta_1$. Perhaps it is differentially regulated and it is up-regulated in different conditions.

Sanchez-Madrid: What about VCAM-1?

Elices: These compounds also inhibit VCAM-1 but they do so at a much lower level—probably about two orders of magnitude lower in terms of *in vitro* activity.

Hynes: Your adoptive transfer blocking experiments weren't very effective: could that be because you were trying to block an interaction with VCAM-1 rather than with fibronectin?

Elices: We and others have shown that in the case of the adoptive delayed-type hypersensitivity, only about 50% of the response is mediated by $\alpha_4\beta_1$ (see references in Table 1). When you inhibit the response with monoclonal antibodies to $\alpha_4\beta_1$, you get about 50% inhibition.

Etzioni: Does your peptide inhibitor have other effects apart from binding to $\alpha_4\beta_1$? We think that the asthma reaction has two phases, an early and a late response. The late response is an inflammatory response compatible with neutrophils arriving, whereas the early response is a consequence of histamine release. You mentioned that your peptide causes an improvement—not much in the late response but definitely in the early response. How do you explain this in the light of the fact that in this phase only histamine takes part, with no effect of adhesion molecules?

Elices: The CS-1 blocker can actually have two effects (this has not yet been fully studied and addressed). One could be an activation of mast cells or other cells that release histamines (these are involved in the early response as you pointed out). A potential explanation is that the peptide, because of its size, has the ability to penetrate the lung tissue, so it could prevent some of the resting mast cells in the lung from being activated. In addition, it is possible that this compound blocks eosinophil adhesion.

Haskard: You mentioned experiments on eosinophil migration *in vitro* and *in vivo* (Kuijpers et al 1993). In the *in vitro* transwell system that you used, you did the experiments in the presence of anti-β_2 antibody. How much inhibition was there when you didn't put the anti-β_2 in?

Elices: $\alpha_4\beta_1$ accounted for 50–70% of the transmigration of the eosinophils across cultured endothelial cells, using C5a or platelet-aggregating factor as a chemoattractant. So it accounts for at least half of the transmigration of the eosinophils, but by no means is it the only mechanism involved in the eosinophil migration.

Hogg: It has been reported that binding to ICAM-1 produces cell flattening but that cells which bind to VCAM-1 only adhere and don't actually flatten (Beekhuizen et al 1992). Do you know whether the interaction with the CS-1 variant of fibronectin is different? Does it support flattening of the eosinophils? This might be relevant to the issue of what is responsible for the transmigration you see. Have you looked to see whether the CS-1-containing fibronectin is produced in other inflammatory conditions?

Elices: We've looked at other inflammatory conditions and we find that CS-1 fibronectin is up-regulated in some but not all of them. In particular, we have found up-regulation of CS-1 fibronectin in the skin and the gut. Inevitably, we find this variant of fibronectin expressed on the endothelium of inflamed tissue.

Hogg: Is it a common response to an inflammatory stimulus or is it unique to rheumatoid arthritis?

Elices: I don't think it's unique to rheumatoid arthritis, but I wouldn't go so far as to say that it is a common thread through all inflammatory reactions.

Hynes: It seems to me that, as much as anything, the effect of inflammatory cytokines is one of relocation, because that form of fibronectin is widely expressed in the basement membranes of endothelial cells.

Pober: In vivo, it's very difficult to find evidence of much fibronectin in basement membranes of endothelial cells in normal settings; but as Bob Colvin showed years ago in rat–mouse skin transplants (Clark et al 1982, and which has been subsequently confirmed by Steve Albelda in human–mouse transplants, Juhasz et al 1993), if you produce a wound such that you get an angiogenic response, fibronectin production increases quite dramatically. In the new vessels, you can see fibronectin clearly in the endothelial basement membrane. As the new vessels mature and form more normal vessels, fibronectin staining disappears again.

Hynes: It goes down: it doesn't disappear. Fibronectin is present in many endothelial basement membranes.

Pober: I suspect that if your test is sensitive enough, you will continue to detect some fibronectin in the basement membrane of mature vessels, but it becomes a relatively uncommon component compared with what you see in the midst of an angiogenic response.

One question about the cytokine effects *in vitro* has to do with cell density. In unpublished endothelial cell experiments, we found that cell density is one of the most dramatic regulators of fibronectin synthesis. Fully confluent endothelial cultures essentially shut down their synthesis of fibronectin and collagen, whereas subconfluent cells produce these molecules copiously. Therefore, when you are looking for a cytokine effect on synthesis of these proteins, it is important to determine whether you are looking at a secondary cell density effect or a direct cytokine effect, because all of these inflammatory cytokines are cytostatic and reduce cell density. Anything you do to produce a less dense culture will markedly increase production of these matrix molecules.

Hynes: I don't know about the eosinophils, but CS-1 fibronectin certainly promotes spreading of lymphoid cells (Guan & Hynes 1990).

Elices: Yes; in our experiments with lymphocytes, above a certain density of CS-1 fibronectin we do see spreading.

Sonnenberg: How is the fibronectin bound to the apical surface of leukocytes? Is this also integrin mediated? If so, it should be possible to block transmigration of leukocytes with RGD peptides, because this would lead to stripping off of the fibronectin from the apical surface.

Elices: That's a very good point. We haven't tried using RGD peptides, nor have we investigated how the fibronectin is bound to the apical surface.

Ruggeri: Your hypothesis of the inhibitory mechanisms of these compounds assumes that fibronectin is located on the outer aspect of the membrane on the luminal surface. Have you proved this is the case?

Elices: Using transmission electron microscopy, we have shown that CS-1 fibronectin is exposed on the luminal surface of the membrane (unpublished results).

Ruggeri: If CS-1 fibronectin occurs on the surface of endothelial cells, can it support other adhesive interactions? I'm thinking in particular about the link between inflammation and activation of haemostasis. Do you have any evidence that fibronectin could support interaction of endothelial cells with platelets, not just as a surface for creating thrombi, but as a surface for assembly of fibrin deposition, which is usually seen in these conditions?

Elices: In rheumatoid arthritis, you see some fibrin deposition in the synovial tissue. We looked at this fibrin deposition using immunohistochemistry and found that it was mostly confined to the synovial membrane (Elices et al 1994). We didn't find very much on the endothelium of the blood vessels. Our conclusion was that CS-1 fibronectin does not provide a surface for fibrin deposition in rheumatoid arthritis.

Ruggeri: It's a complex issue, of course. I think fibrin is probably pathogenetically relevant because it can generate fibrin degradation products that have a very significant pro-inflammatory activity.

There are two different mechanisms that support eosinophil interaction and transmigration in these diseases. Is there any rationale for choosing one approach and blocking one pathway as opposed to the other? Are the two pathways really equivalent? Do you need to block both of them to have an effective therapeutic intervention?

Elices: It's not clear whether *in vivo* you need to block activation or adhesion, or both. One of the advantages of using the small molecular weight inhibitors that we have developed is that you can potentially inhibit both adhesion and activation.

Haskard: If you look at the histology of rheumatoid arthritis, what appears to be hyperplasia of the synovial lining isn't really hyperplasia at all: it is mainly related to the recruitment of bone marrow monocytes (Edwards & Willoughby 1982). There is a school of thought among rheumatologists that monocytes are particularly important in the chronic inflammatory response in rheumatoid arthritis. Do you have any data on blocking the monocyte function or monocyte adhesion?

Elices: No. Most of our studies have been done with lymphocytes or eosinophils. Given the fact that $\alpha_4\beta_1$ is expressed on monocytes, one would presume that blocking it would have an inhibitory effect on these cells also.

Haskard: Monocytes are naturally more adhesive than T cells, so it may be more difficult to block their adhesive function.

Winn: Francisco Sanchez-Madrid asked about blocking both VCAM-1 and CS-1 with your peptide. What dose did you use in your delayed-type hypersensitivity reaction? At high doses you might not be able to distinguish between blocking CS-1 and VCAM-1.

Elices: We started at 600 mg/kg, which is a fairly high dose, but we titrated it down to about 6 mg/kg. This low dose is unlikely to have an effect on $\alpha_4\beta_1$ binding to VCAM-1 in delayed-type hypersensitivity.

Winn: Can you tell the difference between CS-1 or VCAM at these concentrations of monoclonal antibody?

Elices: I can't tell the difference for sure. *In vitro*, those concentrations would not inhibit VCAM-1 very well.

Winn: Eosinophils have both β_2 integrins and $\alpha_4\beta_1$ on their surface yet you have shown virtually complete inhibition of eosinophil emigration just by blocking $\alpha_4\beta_1$. I don't understand why β_2 integrin is not important in eosinophil emigration.

Elices: We and others have shown that $\alpha_4\beta_1$ is important in asthma (Ridger et al 1993, Metzger et al 1994, Abraham et al 1994, Rabb et al 1994). I don't know if β_2 is also important.

Rothlein: Presumably, you gave your CS-1 blocker nebulized, so it's hard to believe that it's going to block the traffic of eosinophils from the vasculature into the alveolar space. Therefore, if you do have a reduced eosinophil count, one has to question whether you are blocking adhesion and migration or whether you are blocking the inflammatory stimuli that elicit them.

Elices: I agree that it's difficult to believe there would be effects on the alveolar migration. I think it's perfectly compatible that these molecules could penetrate into the lung tissue and have systemic effects.

Rothlein: But one gets similar results with antibodies to $\alpha_4\beta_1$. If you give antibodies as an inhalant to sheep, they block the eosinophil influx. We found the same thing with anti-ICAM-1: these antibodies clearly do not block the trafficking, they block the chemokines that are calling these cells in. That's why you see an overall reduced inflammation.

Elices: If I recall correctly, the studies with the anti-$\alpha_4\beta_1$ antibodies don't show a significant difference in the eosinophil count that we see with administration of our CS-1 blocker (Abraham et al 1994). It's perfectly possible that the CS-1 blocker, which is of low molecular weight, penetrates through the lung tissue into the systemic circulation, thus preventing migration of inflammatory cells into the tissue. That would explain why you see a reduced eosinophils count in the broncheoalveolar lavage.

Wagner: Did you look to see if the peptide becomes systemic after it's inhaled?

Elices: No, we haven't been able to do that because at the time we did these experiments we didn't have a suitably labelled compound. We are presently pursuing this. We are also looking into ways to administer the compound systemically. If our hypothesis is correct, when we do this we should see the same protection.

Shaltiel: The strategy of using a series of peptides to determine the minimum structural requirements for recognition, and then modifying this minimum structure to endow it with increased rigidity or lowered susceptibilty to proteolysis, may dramatically change its membrane binding and penetration properties. This may result in different or additional localizations of the peptide and therefore side reactions due to the exposure of the peptide to interactions with additional targets.

Hynes: In this case, everything is probably going on outside the cell, so this may not be such a problem.

Humphries: As I'll show tomorrow, VCAM-1 and fibronectin use similar motifs bind to $\alpha_4\beta_1$; this explains why your agent blocks both. However, VCAM-1 binds approximately three- to fourfold tighter than fibronectin and that's probably the reason why you see better inhibition of fibronectin than VCAM-1. I don't know of an example where a peptidic agent that blocks one receptor can distinguish between different ligands of that receptor.

References

Abraham WM, Sielczak MW, Ahmed A et al 1994 α_4 integrins mediate antigen-induced late bronchial responses and prolonged airway hyperresponsiveness in sheep. J Clin Invest 93:776–787

Beekhuizen H, Verdegaal EME, Blokland I, Van Funtl R 1992 Contribution of ICAM-1 and VCAM-1 to the morphological changes in monocytes bound to human venous endothelial cells stimulated with recombinant interleukin 4 (rIL-4) or IL-1α. Immunology 77:469–472

Clark RA, Quinn JH, Winn HJ, Lanigan JM, Dellepella P, Colvin RB 1982 Fibronectin is produced by blood vessels in response to injury. J Exp Med 156:646–651

Edwards JCW, Willoughby DA 1982 Demonstration of bone marrow derived cells in synovial lining by means of giant intracellular granules as genetic markers. Ann Rheum Dis 41:177–182

Elices MJ, Tsai V, Strahl D et al 1994 Expression and functional significance of alternatively spliced CS1 fibronectin in rheumatoid arthritis microvasculature. J Clin Invest 93:405–416

Guan J-L, Hynes RO 1990 Lymphoid cells recognize an alternatively spliced segment of fibronectin via the integrin receptor $\alpha_4\beta_1$. Cell 60:51–63

Juhasz I, Murphy GF, Yan H-C, Herlyn M, Albelda SM 1993 Regulation of extracellular matrix proteins and integrin cell substratum adhesion receptors on epithelium during cutaneous human wound healing *in vivo*. Am J Pathol 193:1458–1469

Kuijpers TW, Mul EPJ, Blom M et al 1993 Freezing adhesion molecules in a state of high-avidity binding blocks eosinophil migration. J Exp Med 178:279–284

Metzger WJ, Ridger V, Tollefson V, Arrhenius T, Gaeta FCA, Elices M 1994 Anti-VLA-4 antibody and CS-1 peptide inhibitor modifies airway inflammation and bronchial airway hyperresponsiveness (BHR) in the allergic rabbit. J Allergy Clin Immunol 93:125 (abstr)

Rabb HA, Olivenstein R, Issekutz TB, Renzi PM, Martin JG 1994 The role of the leukocyte adhesion molecues VLA-4, LFA-1 and Mac-1 in allergic airway responses in the rat. Am J Respir Crit Care Med 149:1186–1191

Ridger V, Tollefson V, Elices M, Metzger WJ 1993 Anti-VLA-4 antibody modifies airway inflammation and bronchial airway hyperresponsiveness in the allergic rabbit. Proceedings of the 11th Annual Aspen Allergy Conference, Aspen, CO, July 1993

Wayner EA, Garcia-Pardo A, Humphries MJ, McDonald JA, Carter WG 1989 Identification and characterization of the T lymphocyte adhesion receptor for an alternative cell attachment domain (CS-1) in plasma fibronectin. J Cell Biol 109:1321–1330

Adhesion molecules in cutaneous inflammation

Jonathan N. W. N. Barker

St John's Institute of Dermatology, United Medical and Dental Schools, St Thomas' Hospital, Lambeth Palace Road, London SE1 7EH, UK

Abstract. As in other organs, leukocyte adhesion molecules and their ligands play a major role in cutaneous inflammatory events both by directing leukocyte trafficking and by their effects on antigen presentation. Skin biopsies of inflamed skin from patients with diseases such as psoriasis or atopic dermatitis reveal up-regulation of endothelial cell expression of P- and E-selectin, vascular cell adhesion molecule 1 and intercellular adhesion molecule 1. Studies of evolving lesions following UVB irradiation, Mantoux reaction or application of contact allergen, demonstrate that expression of these adhesion molecules parallels leukocyte infiltration into skin. When cutaneous inflammation is widespread (e.g. in erythroderma), soluble forms of these molecules are detectable in serum. *In vitro* studies predict that peptide mediators are important regulatory factors for endothelial adhesion molecules. Intradermal injection of the cytokines interleukin 1, tumour necrosis factor α and interferon γ into normal human skin leads to induction of endothelial adhesion molecules with concomitant infiltration of leukocytes. In addition, neuropeptides rapidly induce P-selectin translocation to the cell membrane and expression of E-selectin. Adhesion molecules also play a crucial role as accessory molecules in the presentation of antigen to T lymphocytes by Langerhans' cells. Expression of selectin ligands by Langerhans' cells is up-regulated by various inflammatory stimuli, suggesting that adhesion molecules may be important in Langerhans' cell migration. The skin, because of its accessibility, is an ideal organ in which to study expression of adhesion molecules and their relationship to inflammatory events. Inflammatory skin diseases are common and inhibition of lymphocyte accumulation in skin is likely to prove of great therapeutic benefit.

1995 Cell adhesion and human disease. Wiley, Chichester (Ciba Foundation Symposium 189) p 91–106

Inflammatory diseases of the skin form a large part of the clinical workload of practising dermatologists. Those diseases that are particularly problematic, including psoriasis and atopic dermatitis, are characterized by the accumulation of T lymphocytes within the dermal and epidermal compartments of the skin. These diseases are problematic for two reasons. Firstly, they are very common: psoriasis affects approximately 2% of the general population of the developed world and atopic dermatitis affects 15% of all children. Secondly, treatments for these diseases are frequently unsatisfactory and are either unacceptable to the patient or are a cause of either localized or systemic toxic side effects. Many studies indicate that T lymphocytes play a fundamental role in the patho-aetiology of these diseases. Immunohistochemical studies demonstrate that T lymphocytes are the predominant infiltrating cell and treatment with either cyclosporin A or FK506, agents which specifically inhibit T cell activation and proliferation, leads to resolution of the disease processes (Barker 1991). It is highly likely, therefore, that elucidation of mechanisms of T cell recruitment into skin and their *in situ* effector functions will lead to identification of novel therapeutic targets.

The skin is the largest organ in the body and, like many other organs such as gut, kidney and lung, is characterized by the close apposition of epithelial and mesenchymal tissue. In contrast to other organs, however, skin is easily obtainable for analysis using a variety of low morbidity techniques, including analysis of scale, suction blister fluid and whole skin biopsy. Because many of the mechanisms which lead to lymphocyte recruitment in the skin are likely to be common to multiple organs, the skin provides an excellent model for examining the mechanisms leading to lymphocyte recruitment at sites of inflammation in humans. A parallel of this might be, for example, the use of skin prick testing to assess causative allergens in bronchial asthma. Furthermore, potential inhibitors of inflammation can be applied directly to the skin, eliminating the need for systemic administration.

Our studies have been designed to investigate *in vivo* mechanisms of cutaneous inflammation in humans of particular relevance to chronic inflammatory skin diseases such as psoriasis and atopic dermatitis. Most specifically, our investigations have concentrated on determining the role of adhesion molecules in lymphocyte recruitment into skin and the role of these molecules in lymphocyte interactions with antigen-presenting cells.

Adhesion molecule expression in normal and inflamed human skin

It has been proposed by Nickoloff (1988) that accumulation of leukocytes into skin involves three distinct phases; recruitment, retention and return to circulation. The initial step in recruitment into inflamed skin is the interaction of leukocytes with post-capillary dermal venular endothelium. In chronic inflammatory dermatoses (for example psoriasis, atopic dermatitis and lichen planus), immunohistochemical staining of frozen sections of whole skin biopsies clearly demonstrates induction of E-selectin and vascular cell adhesion molecule 1 (VCAM-1) on vascular endothelial cells and up-regulation of constitutive intercellular adhesion molecule 1 (ICAM-1) expression. In psoriasis, the most studied disease, dermal post-capillary venules ultrastructurally resemble high endothelial venules of peripheral lymph nodes. Furthermore, using modified Stamper–Woodruff frozen-section adherence assays, psoriatic skin supports adherence of activated peripheral blood lymphocytes, whereas normal skin does not (Chin et al 1990).

Analysis of human skin after experimental induction of various inflammatory stimuli allows the dynamics of adhesion molecule expression to be assessed *in vivo*. Three particular models that have been used are: (a) allergic contact dermatitis (epidermal delayed-type hypersensitivity reaction (Griffiths et al 1991); (b) Mantoux reaction (dermal delayed-type hypersensitivity reaction; and (c) UVB-induced erythema (Norris et al 1991). Elicitation of poison ivy/oak (rhus) dermatitis leads to induction of E-selectin and VCAM-1 and up-regulation of ICAM-1 at 8 h, and more intensive expression of each adhesion molecule at 24–48 h, paralleling infiltration by leukocytes. Following two minimal erythema doses of UVB radiation, E-selectin is induced by 6 h, is maximal at 24 h and has returned almost to basal levels by 72 h. VCAM-1 and ICAM-1 expression is not affected. In contrast, elicitation of Mantoux reactions induces expression of E-selectin and VCAM-1 by 6 h, with maximal expression persisting to 72 h. These studies highlight two important points: firstly, adhesion molecule expression temporally parallels the inflammatory infiltrate and, secondly, the distribution and dynamics of expression vary between inflammatory stimuli. *In vivo* functional experiments in animals demonstrate that cutaneous delayed-type hypersensitivity requires pathways mediated by both $\alpha_L\beta_2$ (LFA-1) and $\alpha_4\beta_1$ (VLA-4) (the ligands for ICAM-1 and VCAM-1 respectively) (Issekutz 1993) and selectin-mediated events. Since several cytokines including tumour necrosis factor (TNF)-α are present in inflamed skin (Ettahadi et al 1994), it is likely that alterations in vascular adhesion molecule expression are mediated via cytokines produced in skin in response to differing inflammatory stimuli (see below).

Although the studies documented above detail expression of vascular adhesion molecules that may be involved in lymphocyte recruitment in many organs, it is apparent from animal studies that a proportion of circulating lymphocytes show a tissue-specific pattern of recirculation. Lymphocytes displaying 'tropism' for skin are almost exclusively of the memory-type subset but can be differentiated from gut-homing memory T cells by their adhesion molecule profile (Mackay et al 1992). This suggests that homing is dependent upon distinct patterns of expression of adhesion molecules by circulating lymphocytes. In humans it has been postulated that skin-homing memory T cells can be separated by their expression of a carbohydrate E-selectin ligand, cutaneous lymphocyte antigen (CLA), recognized by monoclonal antibody HECA-452 (Picker et al 1991), although functional studies to determine whether these T cells bind more avidly to dermal vascular endothelium have not been performed. Psoriasis, when associated with arthritis, provides an ideal model to establish whether CLA is disease or organ specific. We have, therefore, simultaneously examined CLA expression in lesional skin, synovial fluid and peripheral blood of psoriatic patients with active skin and joint disease, by immunohistology and fluorescence-activated cell sorter analysis. In skin, 47–69% of CD3-positive cells expressed CLA, particularly at the tips of the dermal papillae, whereas the percentage range was 8–13% in synovial fluid and 2–10% in peripheral blood (Smith et al 1994). These data examining a single disease manifested in two organs support the concept of skin-specific lymphocyte homing in humans and indicate that these cells can be identified by CLA expression. Further confirmation, however, is required from the examination of T cell populations in biopsies of synovial membrane.

Regulation of adhesion molecule expression *in vivo*

In vivo, the cytokines interleukin (IL)-1, TNF-α, and γ-interferon play a major role in the induction of vascular endothelial adhesion molecules. Furthermore, studies suggest they play a critial role in cutaneous inflammatory processes and in the pathogenesis of chronic inflammatory cutaneous diseases. Our experimental approach has been to determine the local effects of these cytokines administered intradermally into normal volunteers and observe induced changes by immunohistochemistry of whole skin biopsies (Barker et al 1990). While each of the above cytokines can be shown to be pro-inflammatory *in vivo*, the changes observed may vary from those predicted by *in vitro* studies. For example, intradermal injection of TNF-α results in a biophasic response (R. W. Groves, personal communication). Six hours after a single injection, neutrophils predominate, whilst repeated injections over 6 d induce sustained E-selectin, VCAM-1 and ICAM-1 expression, together with an associated lymphocytic infiltration. As these are human studies, the number and timing of the biopsies are limited, as is the number of cytokines available for us to study. No studies have examined responses in humans to multiple cytokines, which *in vitro* appear to have marked effects on modulation of vascular adhesion molecules. However, in baboon skin, IL-4 injected concomitantly with TNF-α significantly increases TNF-α-mediated VCAM-1 expression and dermal T cell infiltration (Briscoe et al 1992).

Neuropeptides have recently been recognized as potential mediators of cutaneous inflammation and have been implicated in the pathogenesis of several skin diseases, including atopic dermatitis and psoriasis. We have therefore examined modulation of endothelial adhesion molecule expression *in vivo* (Smith et al 1993). Intradermal injection of sensory peptides, substance P, vasoactive intestinal polypeptide (VIP) and calcitonin gene-related peptide (CGRP) is accompanied by weal and flare formation and a rapid infiltration into the dermis of neutrophils. These changes are paralleled by rapid translocation of P-selectin from cytoplasmic Weibel–Palade bodies to endothelial luminal surface and induction of E-selectin expression by 4 h. Interestingly, in some volunteers, an eosinophil-rich infiltrate was observed in the absence of changes in VCAM-1 expression.

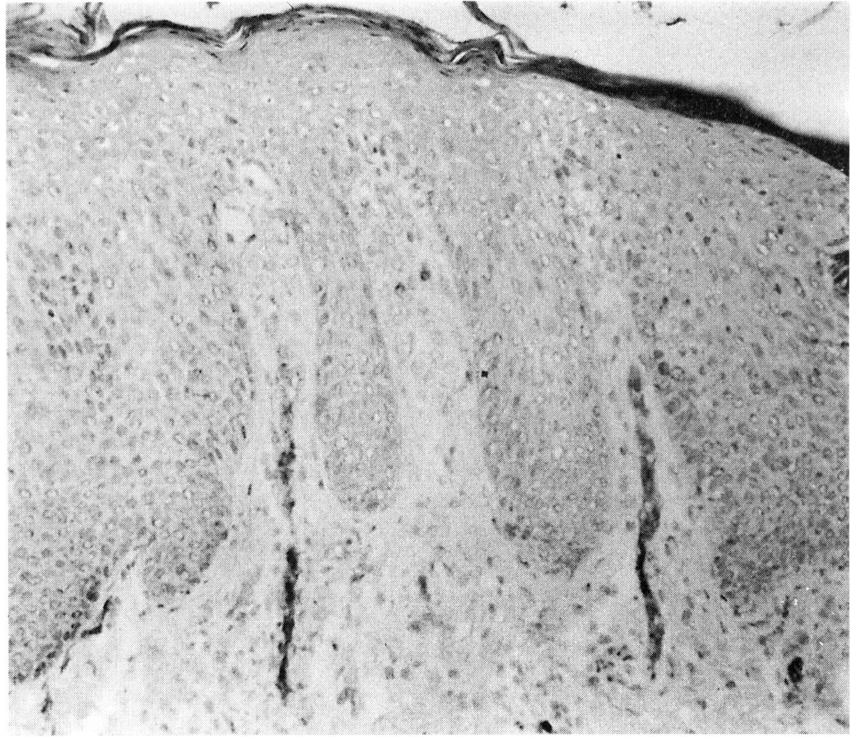

FIG. 1. Photomicrograph of a section of human psoriatic skin stained for E-selectin. Strong expression is observed on dermal papillary blood vessels. Expression of E-selectin is not seen in normal skin. Magnification × 100.

Virtually all resident cell types (including bone marrow-derived cells, fibroblasts and nerve fibres in the dermis, and keratinocytes, Langerhans' cells and melanocytes in the epidermis) synthesize and release mediators capable of activating endothelium, including peptides (e.g. cytokines), lipid and preformed mediators (histamine), and complement factors. The relative role of each cell type and mediator is likely to depend upon the underlying pathophysiology of the disease or, in the case of injury or infection, the source and type of initiating agent. Evidence is accumulating that factors that inhibit or down-regulate inflammatory responses may also be produced by skin-derived cells and induced by various inflammatory stimuli. For example, IL-10 can be detected in skin following UVB irradiation.

In the chronic inflammatory dermatoses (e.g. psoriasis, atopic dermatitis) most inflammatory events occur within dermal papillae, where changes in adhesion molecule expression are most pronounced. In such circumstances, keratinocytes are uniquely situated to provide stimuli required for endothelial cell activation. In a spatial alignment analogous to that of the Bowman's capsule of renal glomeruli, interactions between keratinocytes and endothelial cells mediated via production of cytokines and adhesion molecules are likely to provide the molecular basis for the 'squirting papilla' of Pinckus and Mehregan. This is a morphological description of the site of leukocyte migration into the epidermis (Barker et al 1991) observed in psoriasis.

Circulating adhesion molecules in skin disease

Soluble forms of the selectins (E, L, P), ICAM-1 and VCAM-1 are produced by many cell types *in vitro*, including mononuclear cells, endothelial cells, keratinocytes and melanocytes (Gearing & Newman 1993). While the function of these soluble forms remains unknown, it is possible that they contribute to disease pathology in a variety of ways. For example, soluble adhesion molecules may interfere with normal cell trafficking by inhibiting leukocyte binding to vascular endothelium through competitive blocking of receptors.

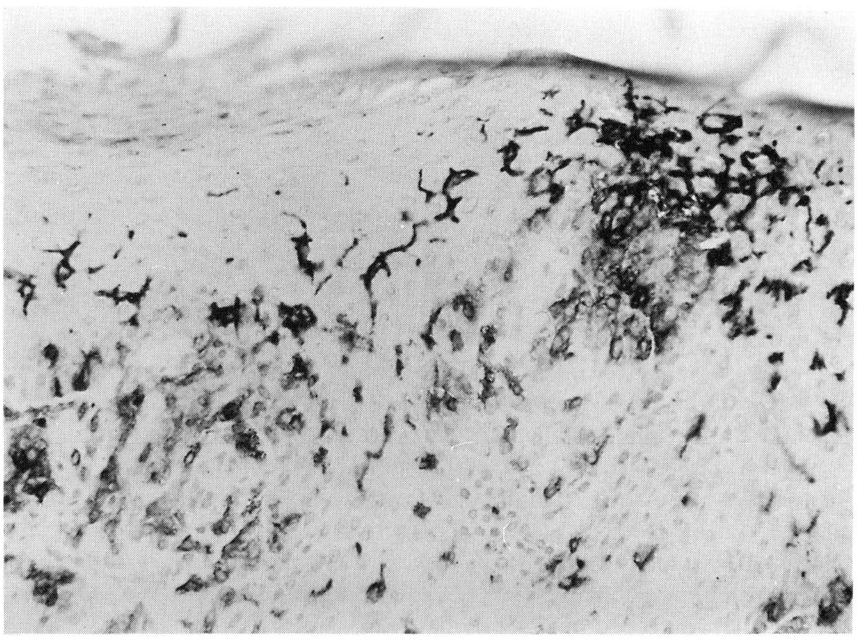

FIG. 2. Photomicrograph of psoriatic skin stained for sialyl Lewis X. Strong expression is seen on epidermal dendritic cells, namely Langerhans' cells, particularly at the tips of dermal papillae. Magnification × 150.

Similarly, binding to ligand receptors may trigger effector responses in the target cell.

Regardless of their pathogenic physiological role, assessment of circulating levels of these adhesion molecules may provide a useful diagnostic or prognostic tool for a wide range of inflammatory diseases. To date, elevated levels of these adhesion molecules have been frequently found in many non-cutaneous diseases. With the exception of studies measuring circulating ICAM-1 in psoriasis (Kowalzicki et al 1993), there are presently no data on the presence of circulating ICAM-1, VCAM-1 and E-selectin in inflammatory skin diseases. An objective, surrogate measure of disease severity would be particularly useful in these diseases, as there are no reliable biochemical parameters of disease activity and clinical scoring measurements do not necessarily accurately reflect activity.

Erythroderma is a serious life-threatening skin disease characterized clinically by generalized erythema and scaling of the skin. It may be a complication of psoriasis, eczema or other related diseases. As a preliminary study to evaluate the use of measuring circulating adhesion molecules in inflammatory skin disease, we recruited 14 patients with erythroderma and assayed circulating forms of adhesion molecules by ELISA (enzyme-linked immunosorbent assay) in their peripheral blood. Circulating ICAM-1, VCAM-1 and E-selectin were all significantly elevated in these patients compared with controls. Circulating ICAM-1 and E-selectin correlated significantly with each other in these patients but circulating VCAM-1 did not. When we grouped the patients according to the underlying cause of the erythroderma, no significant differences between the groups could be detected (Groves et al 1992).

This study is presently being extended to the investigation of less widespread forms of atopic eczema and psoriasis, and also to observe whether levels of circulating adhesion molecules parallel clinical response of these diseases to therapy.

Adhesion molecules in Langerhans' cells

Langerhans' cells are bone marrow-derived dendritic cells of monocyte/macrophage lineage which normally reside in the epidermal compartment of skin (Stingl et al 1989). In human epidermis, Langerhans' cells are characterized ultrastructurally by the presence of cytoplasmic Birbeck granules and phenotypically by surface expression of CD1a and class II major histocompatibility complex antigens. Functional *in vitro* studies show that isolated Langerhans' cells can process and subsequently present antigen to T lymphocytes, while *in vivo* studies demonstrate a critical role for Langerhans' cells in the generation of cell-mediated immune responses in the skin. For example, UVB irradiation of intact human skin, which depletes the epidermis of Langerhans' cells, abrogates cutaneous responses to topically applied antigen.

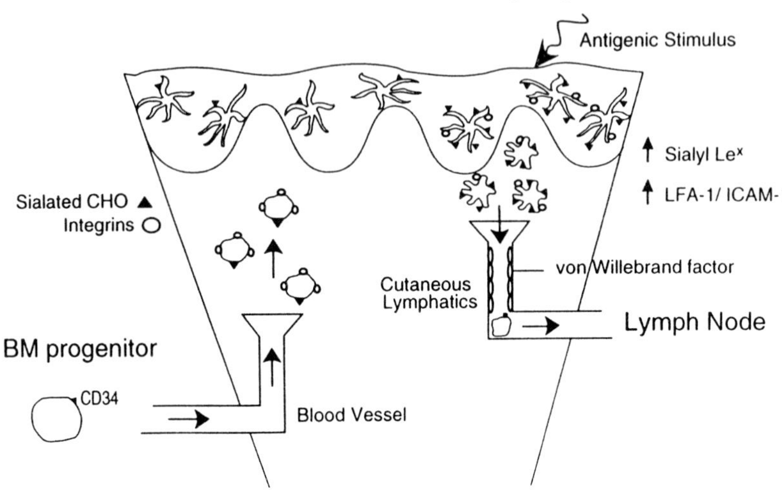

FIG. 3. Schematic diagram representing our working hypothesis of the role of adhesion molecules in Langerhans' cell trafficking (courtesy of E. L. Ross). CD34-positive bone marrow-derived Langerhans' cell (BM) progenitors circulate into skin via the dermal vasculature and migrate into the epidermis. After antigenic stimulation, Langerhans' cells leave the epidermis and migrate to regional lymph nodes via afferent lymphatics. This is accompanied by up-regulation of carbohydrate selectin ligands and β_2 integrins.

Furthermore, a central role for Langerhans' cells in the pathogenesis of the inflammatory dermatoses, psoriasis and atopic dermatitis, has been suggested on the basis of the fact that Langerhans' cells isolated from the epidermis of these diseases are potent antigen-presenting cells when compared with normal skin (Baadsgaard et al 1990).

It is also known that Langerhans' cells are migratory. Growth and maturation in bone marrow is followed by their movement into the skin, presumably through post-capillary venules. After exposure to antigen, Langerhans' cells leave the epidermis and migrate to regional lymph nodes via afferent lymphatics.

It is likely, therefore, that adhesion molecules play a fundamental role in Langerhans' cell biology through their influence on Langerhans' cell trafficking and their role as accessory molecules in antigen presentation to T lymphocytes. Antigen presentation can be assessed *in vitro* using the mixed epidermal cell lymphocyte reaction (MELR). In this assay, proliferation of allogeneic T lymphocytes can be inhibited by the addition of antibodies to ICAM-1 or LFA-1. Fluorescence-activated cell sorter analysis of Langerhans' cell suspensions demonstrates expression of both ICAM-1 and LFA-1 (Simon et al 1991). *In vivo*, these molecules are not expressed by Langerhans' cells in normal skin. Interestingly, however, in both psoriasis and atopic dermatitis, where Langerhans' cells are more potent antigen-presenting cells, these molecules are expressed *in vivo* (McGregor et al 1992). ICAM-3, a further accessory molecule important in antigen presentation, is also expressed by Langerhans' cells *in vivo* (Acevedo et al 1993).

Because both selectin- and integrin-mediated pathways are required in sequence for leukocyte trafficking, it is likely that similar mechanisms apply to Langerhans' cell migration. We have investigated Langerhans' cell expression of selectin ligands *in vivo* by immunohistology and also in short-term organ culture experiments (Ross et al 1994). These studies show that Langerhans' cells up-regulate expression of sialyl Lewis X in parallel to migration from the epidermis after epicutaneous contact with allergen in presensitized individuals. Concomitantly, E-selectin is up-regulated on dermal papillary blood vessels. Since Langerhans' cells gain access to peripheral lymph nodes by migration along afferent lymphatic channels, it is attractive to speculate that events mediated by adhesion molecules control this process. Although it is not known whether lymphatic endothelium expresses leukocyte adhesion molecules, cultured lymphatic endothelial cells from both rats and humans contain von Willebrand factor, which in vascular endothelium co-localizes with P-selectin in cytoplasmic Weibel–Palade bodies (Djoneidi & Brodt 1991).

Adhesion molecules and therapy for inflamed skin

Our increasingly detailed knowledge of adhesion molecules and the mechanisms which control leukocyte trafficking suggests that molecules aimed at inhibiting their functions may be useful therapeutic agents. Development of treatments aimed at inhibiting T lymphocyte accumulation in skin, perhaps by inhibition of skin-specific pathways, is likely to prove of great therapeutic benefit in inflammatory dermatoses. The skin is an immunologically privileged site and, from an investigative and therapeutic point of view, it has the benefit that agents of interest may have activity when applied directly to its surface. This is particularly the case for small molecules which can penetrate the epidermal barrier. For this reason, carbohydrates that block selectin-mediated adhesion pathways remain an attractive option for future dermatological therapy.

Acknowledgements

The author would like to thank the many collaborators on whose work this text is based, particularly, Michael Allen, Elizabeth Ross, Catherine Smith, Richard Groves, Donald MacDonald, Chrisopher Griffiths, Brian Nickoloff and Dorian Haskard. The original work presented in this manuscript has been supported by the Wellcome Trust and Dunhill Medical Trust.

References

Acevedo A, Del Pozo MA, Arroyo AG, Sanchez-Mateos P, Gonzales-Amaro R, Sanchez-Madrid F 1993 Distribution of ICAM-3-bearing cells in normal human tissues. Am J Pathol 143:774–783

Baadsgaard O, Fisher GJ, Voorhees JJ, Cooper KD 1990 Interactions of epidermal cells and T cells in inflammatory skin diseases. J Am Acad Dermatol 233:1312–1317

Barker JNWN 1991 Pathophysiology of psoriasis. Lancet 338:227–230

Barker JNWN, Allen MH, MacDonald DM 1990 Alterations induced in normal human skin by in-vivo interferon-γ. Br J Dermatol 122:451–458

Barker JNWN, Mitra RS, Griffiths CEM, Dixit VM, Nickoloff BJ 1991 Keratinocytes as initiators of inflammation. Lancet 337:211–214

Briscoe DM, Cotran RS, Pober LS 1992 Effects of tumor necrosis factor, lipopolysaccharide and IL-4 on the expression of vascular cell adhesion molecule in vivo. J Immunol 149:2954–2960

Chin YH, Falanga V, Taylor JR, Cai JP, Bax J 1990 Adhesion of human helper/memory T cell subsets to psoriatic dermal endothelium. J Invest Dermatol 94:413–418

Djoneidi M, Brodt P 1991 Isolation and characterisation of rat lymphatic endothelial cells. Microcirc Endothelium Lymphatics 7:161–181

Ettahadi P, Greaves MW, Wallach D, Aderka D, Camp RDR 1994 Elevated tumour necrosis factor-α biological activity in psoriatic skin lesions. Clin Exp Immunol 96:146–151

Gearing AJH, Newman W 1993 Circulating adhesion molecules in disease. Immunol Today 14:506–512

Griffiths CEM, Barker JNWN, Kunkel S, Nickoloff BJ 1991 Modulation of leucocyte adhesion molecules, a T-cell chemotaxin (IL-8) and a regulatory cytokine (TNF-α) in allergic contact dermatitis (rhus dermatitis). Br J Dermatol 124:519–526

Groves RW, Barker JNWN, Haskard DO, Bird C, MacDonald DM 1992 Circulating cytokines and soluble adhesion molecules in widespread inflammatory skin disease. Br J Dermatol 127:428(abstr)

Issekutz TB 1993 Dual inhibition of VLA-4 and LFA-1 maximally inhibits cutaneous delayed-type hypersensitivity-induced inflammation. Am J Pathol 143:1286–1293

Kowalzicki L, Bildau H, Neuber K, Kohler I, Ring J 1993 Clinical improvement in psoriasis during dithranol/UVB therapy does not correspond with decreased elevation of circulating ICAM-1 levels. Arch Dermatol Res 285:233–235

McGregor JM, Barker JNWN, Ross EL, MacDonald DM 1992 Epidermal dendritic cells in psoriasis possess a phenotype associated with antigen presentation: in situ expression of β_2 integrins. J Am Acad Dermatol 27:383–388

Mackay CR, Marston WL, Dudler L, Spertini O, Tedder TF, Hein WR 1992 Tissue-specific migration pathways by phenotypically distinct populations of memory T cells. Eur J Immunol 22:887–895

Nickoloff BJ 1988 Role of γ-interferon in cutaneous trafficking of lymphocytes with emphasis on molecular and cellular adhesion events. Arch Dermatol 124:1835–1843

Norris P, Poston RN, Thomas DS, Thornhill M, Hawk J, Haskard DO 1991 Expression of ELAM-1, ICAM-1 and VCAM-1 in experimentally induced cutaneous inflammation: a comparison of ultraviolet B erythema and delayed hypersensitivity. J Invest Dermatol 96:763–770

Picker LJ, Kishimoto TK, Smith CW, Warnock RA, Butcher EC 1991 ELAM-1 is an adhesion molecule for skin-homing T cells. Nature 349:796–799

Ross EL, Barker JNWN, Allen MH, Chu AC, Groves RW, MacDonald DM 1994 Langerhans' cell expression of the selectin ligand, sialyl Lewis X. Immunol 81:303–308

Simon JC, Cruz PD, Tigelaar RE, Sontheimer RD, Bergstresser PR 1991 Adhesion molecules CD11a, CD18, and ICAM-1 on human epidermal Langerhans cells serve a functional role in the activation of alloreactive T cells. J Invest Dermatol 96:148–151
Smith CH, Barker JNWN, Morris RW, MacDonald DM, Lee TH 1993 Neuropeptides induce rapid expression of endothelial adhesion molecules and elicit granulocytic infiltration in human skin. J Immunol 151:3274–3282
Smith CH, Pitzalis C, Yanni G, Pipitone N, Corrigall VM, Barker JNWN, Panayi GS 1994 Cutaneous lymphocyte antigen positive lymphocytes selectively infiltrate skin but not joints in psoriasis. J Invest Dermatol 102:531 (abstr)
Stingl G, Tschachler E, Groh V, Wolff K 1989 Immune functions of epidermal cells. In: Norris DA (ed) Immune mechanisms in cutaneous disease. Marcel Dekker, New York, p 3–72

DISCUSSION

Stanley: You pointed out very nicely that in these different dermatological conditions, such as allergic contact dermatitis and psoriasis, you see lymphocytes in the skin. In inflammatory skin disorders, there are a range mechanisms by which the lymphocytes might be getting out of the blood vessels into the skin. Is this a non-specific feature of inflammation? In other words, whenever you get inflammation, does the body want to have lymphocytes going through the skin to see what's going on, acting as a sort of monitor? Or is there anything specific about the role of the lymphocytes in the skin (either in the pathways or in the lymphocytes themselves)? Psoriasis looks very different from allergic contact dermatitis. If lymphocytes are involved in the pathogenesis of psoriasis and they are involved in the pathogenesis of allergic contact dermatitis, they must be doing very different things in the skin.

Barker: I don't think there are any differences in the mechanisms of lymphocyte recruitment in these diseases. The only differences observed so far have been to do with the nature of the lymphocytes: for example, it would appear that lymphocytes in psoriasis are mainly of the Th_1 type (they produce IL-2 and γ-interferon), whereas there's evidence in atopic dermatitis that they are of the Th_2 type (they produce IL-4 and IL-5). Most of the T cells that you find in skin are of the memory phenotype, but I don't think anyone has been able to demonstrate clones in non-malignant skin conditions, including allergic contact dermatitis, where you might expect to be able to detect clones of T cells.

Stanley: The implication is that the lymphocytes are involved in the pathogenesis, because if they're not involved in the pathogenesis, then there might not be differences. There might just be inflammation that causes lymphocytes to come in to the skin.

Barker: I certainly think that they are involved in the pathogenesis. Clearly, in allergic contact dermatitis, at least some of the infiltrating lymphocytes have to be involved in it. All the available evidence for psoriasis and eczema is that they are playing an important role, because if you prevent their function, you at least prevent the expression of the disease. This doesn't mean to say they are the primary abnormality, but they're certainly involved in some way.

Stanley: In the system in which you looked for Langerhans' cells, using short-term organ culture of skin biopsies, the theory is that E-cadherin may be important in keeping the Langerhans' cell within the epidermis, but you didn't show what happens to E-cadherin: is it down-regulated in these cells?

Barker: I didn't mention Mark Udey's work on E-cadherin (Tang et al 1993). This is an elegant mechanism explaining how these cells might stick and what makes them epidermotropic. Until recently we haven't had much of a clue as to how Langerhans' cells remain within the epidermis, but this paper demonstrates that Langerhans' cells express E-cadherin, which of course keratinocytes also express. Homotypic adhesion between Langerhans' cells and keratinocytes via E-cadherin occurred *in vitro* and this is a potential explanation as to why Langerhans' cells remained within the epidermal compartment. Of course, if Langerhans' cells are migrating out of the epidermis, you would expect them to down-regulate their E-cadherin; we haven't looked at this yet.

Pober: Some observations from experiments we've conducted suggest that the regulation of what is going on in these dermal microvessels is more complex. When we were doing injection studies, we found that endotoxin did not induce VCAM-1 and did not induce lymphocyte recruitment in baboon skin, whereas TNF-α did (Briscoe et al 1992). To explore this further, we turned to organ culture (Petzelbauer et al 1993). We found that if we took the biopsy first and then either injected it with or bathed it in cytokines, none of the combinations of cytokines we added induced VCAM-1 expression. Furthermore, we failed to find any expression of any adhesion molecule on the capillary loops that fed the epidermis. We could only induce adhesion molecules—E-selectin primarily—on the post-capillary venules, although ICAM-1 could be expressed on the capillary loops and was up-regulated. We wondered what was controlling this, because clearly VCAM-1 can be expressed on these vessels and, as Jonathan Barker has shown, in psoriasis, the capillary loops express these adhesion molecules.

We've recently done an organ culture study in collaboration with Erwin Braverman at Yale, in which we biopsied non-lesional, perilesional and lesional skin of psoriatics (Petzelbauer et al 1994). Rather than look only at the initial level of expression of adhesion molecules, we also looked at the responsiveness to cytokines in an organ culture bath. Basically, non-lesional skin from psoriatics is similar to normal skin from everyone else: you can only induce E-selectin (and not VCAM-1) and you can only induce E-selectin on the post-capillary venules (not on the capillaries). Although no lymphocytes can be detected immunocytochemically, perilesional skin can express VCAM-1 as well as E-selectin on the post-capillary venules. In lesional skin, you can induce E-selectin and VCAM-1 on the tips of the capillary loops next to the epidermis. The portion of the capillary that is responsive to cytokines is precisely the portion which has changed the morphological appearance of its basement membrane to look like venule basement membrane (laminated) instead of arteriole basement membrane (homogeneous). I don't know whether the basement membrane change is causal, but it seems to be concomitant with the change in the ability of endothelial cells to respond to cytokines. The point is that there are molecules other than cytokines that are regulating inflammatory processes and the skin is not simply a blank substrate in which any cytokine will give you any response. There are factors that are yet to be characterized that will control what kind of response you get.

Barker: That's very interesting. Certainly, microvascular endothelial cells in culture do behave differently from other endothelial cells.

Pober: Also, different microvasculature preparations behave differently. Our dermal microvascular preparations (Petzelbauer et al 1993) appear to be derived from superior vascular plexus cells which we think differ from the ones that Tom Lawley has developed most likely from the deep vascular plexus (Swerlick et al 1992). Our cells have a different pattern of cytokine response to his.

Pals: In our lab, Drs Das and de Boer have done studies comparing contact dermatitis and psoriasis, and they have found clear differences in adhesion molecule expression. In psoriasis they find relatively high levels of E-selectin and almost no VCAM-1 expression, whereas in contact dermatitis there is very high expression of both E-selectin and VCAM-1 on the vessels (Das et al 1994).

Another interesting observation they made was that in non-lesional skin of psoriatics there was up-regulation of adhesion molecules. For instance, on about 50% of Langerhans' cells they found high expression of HECA-452 (cutaneous lymphocyte antigen). In addition there was up-regulation of ICAM-1 and E-selectin on venules (de Boer et al 1994).

Barker: I agree that there are some differences. The problem with comparing delayed-type hypersensitivity and psoriasis directly is that delayed-type hypersensitivity is an early disease (it's all over within 72 h or so), whereas psoriasis is a chronic disease. For example, we've looked for P-selectin in psoriasis; we find that some vessels express it on their luminal membrane and other vessels express it in Weibel–Palade bodies. Presumably, these molecules are being turned on and off repeatedly as a dynamic process.

Hynes: Are adhesion molecules present in lymphatic vessels?

Wagner: There is some evidence that lymphatic endothelium contains von Willebrand factor and likely also P-selectin (Tabuchi & Yamamoto 1974, Magari et al 1989). Perhaps von Willebrand factor has other functions besides haemostasis. It could also be involved in leukocyte recruitment or adhesion, i.e. a function related to inflammation.

Ruggeri: The importance of von Willebrand factor in adhesion has been demonstrated by a knockout experiment of Nature—von Willebrand's disease. In this disease, the only abnormality is defective platelet function and abnormal haemostasis.

Verrando: What is the role of Langerhans' cells in psoriasis?

Barker: The cause of psoriasis is unknown, although it is believed to be immunologically mediated. In lesional psoriatic skin, Langerhans' cells are activated as demonstrated by increased ability to present antigen to T lymphocytes. What the antigen is, however, is unknown. Other lines of evidence suggesting that psoriasis is an immunologically mediated disease are: (a) an association with certain HLA antigens, particularly CW-6; (b) an early influx of T lymphocytes into lesional tissue; and (c) a rapid response of disease to anti-T cell therapy such as cyclosporin A (Barker 1991).

Garrod: You talked about how Langerhans' cells may be recruited into the epidermis and how, after antigen challenge, they migrate from the skin. But how are they retained within the epidermis in normal skin and at what level of the epidermis do they reside?

Barker: There is evidence from studies in mice that epidermal Langerhans' cells express E-cadherin which allows them to bind homotypically to keratinocytes (Tang et al 1993). Maximal expression of E-cadherin is in the mid epidermis, which is also the main epidermal level at which Langerhans' cells reside.

Stanley: What about P-cadherin? P-cadherin is expressed in the basal layer of the epidermis, although its function is unknown.

Haskard: Cytokines may not be the only way in which endothelial cells can be activated to express adhesion molecules such as E-selectin and VCAM-1. We have recently followed up a report by Damle et al (1991) who found that direct contact between T cells and umbilical vein endothelial cells can result in the expression of E-selectin and VCAM-1. We found that co-incubation of fixed T cells and dermal microvascular endothelial cells resulted in E-selectin and VCAM-1 expression and this was not blocked by a cocktail of antibodies to IL-1α, IL-1β, TNF-α and TNF-β (Yarwood et al 1994). We do not know the mechanism for this effect, but the observation suggests that T cells in the skin might have a direct effect on endothelial activation.

Barker: In chemotaxis assays, most cytokines can be demonstrated to possess chemotactic activity, including IL-8. However, in the limited number of experiments where IL-8 has been injected into human skin, the leukocytic infiltrate resulting is much weaker than you would predict from its *in vitro* activity. This probably reflects the fact that IL-8 does not up-regulate adhesion molecules on endothelial cells. A more relevant *in vivo* experiment would therefore be to stimulate endothelial adhesion molecule production and subsequently to inject a chemoattractant/leukocyte activator, such as IL-8.

Hynes: Could you say a little more about why you looked at the effects of neuropeptides on the expression of endothelial adhesion molecules?

Barker: We examined the *in vivo* proinflammatory effects of neuropeptides because they have recently been implicated in inflammatory skin diseases such as atopic eczema and psoriasis (Smith et al 1993). For example, in the skin, increased amounts of neuropeptide can be detected in these conditions and people have thought for a long time that there is a neural component to this disease. The symmetrical nature of the disease is an example of this. To date, it has only been known that neuropeptides produce a weal and flare response accompanied by plasma leakage. We were able to demonstrate that neuropeptides, found in human skin, produce a leukocytic infiltrate when injected into human skin and that this accumulation of leukocytes is accompanied by a rapid up-regulation of vascular cell adhesion molecule expression.

Birchmeier: Do Langerhans' cells undergo a transformation from mesenchymal to epithelial cells?

Barker: Langerhans' cells are bone marrow-derived mesenchymal cells and not epithelial cells. They do, however, reside within the epidermis, presumably as a result of the mechanisms discussed earlier. It is interesting that a T cell line derived from a patient with human cutaneous T cell lymphoma also expresses E-cadherin (Nickoloff et al 1993). These T cells are epidermotropic; perhaps this is the mechanism by which they bind to keratinocytes.

References

Barker JNWN 1991 Pathophysiology of psoriasis. Lancet 338:227–230

Briscoe DM, Cotran RS, Pober JS 1992 Effects of tumor necrosis factor, lipopolysaccharide and IL-4 on the expression of vascular cell adhesion molecule-1 *in vivo*. Correlation with $CD3^+$ T cell infiltration. J Immunol 149:2954–2960

Damle NK, Eberhardt C, Van der Vieren M 1991 Direct interaction with primed $CD4^+$ $CD45RO^+$ memory T lymphocytes induces expression of endothelial leukocyte adhesion molecule-1 and vascular cell adhesion molecule-1 on the surface of vascular endothelial cells. Eur J Immunol 21:2915–2923

Das PK, de Boer OJ, Visser A, Verhagen CE, Bos JD, Pals ST 1994 Differential expression of ICAM-1, E-selectin and VCAM-1 by endothelial cells in psoriasis and contact dermatitis. Acta Dermat Venereol Suppl 186:21–22

de Boer OJ, Wakelkamp IMMJ, Claessen N et al 1994 Increased expression of adhesion molecules in both involved and non-involved psoriatic skin Arch Dermat Res 286:304–311

Magari S, Ito Y, Sakanaka M 1989 An immunoelectron microscopic study of von Willebrand factor in the thoracic duct endothelium of rats. Lymphology 22:76–80

Nickoloff BJ, Trinh DT, Nestle FO, Mitra RS 1993 HUT78 cells express E-cadherin and bind to keratinocytes which also express functional E-cadherin molecules. J Cutaneous Pathol 20:560(abstr)

Petzelbauer P, Bender JR, Wilson J, Pober JS 1993 Heterogeneity of dermal microvascular endothelial cell antigen expression and cytokine responsiveness in situ and in cell culture. J Immunol 151:5062–5072

Petzelbauer P, Pober JS, Keh A, Braverman IM 1994 Inducibility and expression of microvascular endothelial adhesion molecules in lesional, perilesional and uninvolved skin of psoriatic patients. J Invest Dermatol 103:300–305

Smith CH, Barker JNWN, MacDonald DM, Lee TH 1993 Neuropeptides induce rapid expression of endothelial cell adhesion molecules and elicit granulocytic infiltration in human skin. J Immunol 151:3274–3282

Swerlick RA, Lee KH, Li LJ, Sepp NT, Caughman SW, Lawley TJ 1992 Regulation of vascular cell adhesion molecule 1 on human microvasular endothelial cells. J Immunol 149:698–705

Tabuchi H, Yamamoto T 1974 Specific granules in the endothelia of blood and lymphatic vessels in the cardiac valves of dogs. Arch Histol Jpn 37:217–224

Tang A, Amagia M, Granger LGT, Stanley JR, Udey MC 1993 Adhesion of epidermal Langerhans' cells to keratinocytes mediated by E-cadherin. Nature 361:82–85

Yarwood H, Mason JC, Mahiouz D, Sugars K, Haskard DO 1994 Characterisation of E-selectin and vascular cell adhesion molecule-1 (VCAM-1) induction on human umbilical vein endothelial cells and dermal microvascular endothelial cells during co-culture with T-lymphocytes. FASEB (Fed Am Soc Exp Biol) J 8:133(abstr)

Defective cell-cell adhesion in the epidermis

John R. Stanley

Dermatology Branch, National Cancer Institute, Building 10, Room 12N238, Bethesda, MD 20892, USA

Abstract. The disastrous effects of loss of epidermal cell adhesion are epitomized by the life-threatening blistering skin diseases pemphigus foliaceus and pemphigus vulgaris. Clinical and experimental observations show that loss of cell adhesion is induced by these patients' autoantibodies. Pemphigus foliaceus antigen is desmoglein 1 (dsg-1), a desmosomal transmembrane glycoprotein limited in distribution to stratified squamous epithelia. It is linked to plakogoblin, a desmosomal plaque protein. Molecular cloning has shown that desmogleins are members of the cadherin gene superfamily. The originally described cadherins (e.g. E-cadherin) are transmembrane, calcium-dependent, homophilic adhesion molecules. Pemphigus vulgaris antigen is a 130 kDa glycoprotein also linked to plakoglobin. Molecular cloning has shown that pemphigus vulgaris antigen is also a desmoglein, dsg-3. Antibodies against pemphigus vulgaris antigen subdomains homologous to the binding subdomains of classical cadherins cause loss of epidermal cell adhesion, which suggests that desmogleins mediate adhesion, although direct evidence for this is lacking. The extracellular domain of pemphigus vulgaris antigen cannot substitute in function for that of E-cadherin. Future studies should address the cell biological function of desmogleins.

1995 Cell adhesion and human disease. Wiley, Chichester (Ciba Foundation Symposium 189) p 107–123

The recent explosion of interest in, and knowledge about, cell adhesion molecules attests to their importance in many facets of biology. These molecules have proved to be important in normal development and physiology, and in the pathophysiology of carcinogenesis, inflammation, and hereditary and autoimmune diseases.

Classically, in medicine we have learned about normal physiology and cell biology by the study of disease. Skin disease is particularly amenable to study because of the accessibility and visibility of the skin. In this paper I will try to demonstrate how two autoimmune diseases of the skin, pemphigus vulgaris and pemphigus foliaceus, illustrate the devastating consequences of loss of cell–cell adhesion in the epidermis and how these diseases hold clues to the nature and potential function of keratinocyte cell adhesion molecules.

Pemphigus vulgaris and pemphigus foliaceus: diseases of cell adhesion

When keratinocytes in the epidermis lose their normal ability to adhere to each other, fluid (serum or serous fluid) collects in the intercellular spaces. This is seen clinically as a blister. However, because the epidermis becomes very fragile as a consequence of this loss of adhesion, the blister breaks easily and is rapidly replaced by an erosion, often with a crust of dried serous fluid. Therefore, in diseases of cell adhesion, the usual clinical presentation on the skin is crusted erosions, sometimes with intact blisters if early lesions can be found.

In pemphigus vulgaris, lesions occur not only on skin, but also on mucous membranes (Stanley 1990, 1992). Therefore, pemphigus vulgaris is a disease of stratified squamous epithelia. Other epithelia are normal. The disease characteristically presents with lesions in the mouth; almost all patients have some mucous membrane lesions during the course of their disease (Fig. 1a). Mucous membrane lesions are almost always seen as erosions, presumably because the initial blister is so fragile it breaks before it is detected on clinical examination. Skin lesions start as flaccid blisters, but rapidly progress into crusted erosions. Because of the fragility of the skin, these erosions often extend at the edges and can become quite large. In active pemphigus vulgaris, the skin is often so fragile that just rubbing on normal-appearing skin can cause an erosion.

Histopathology reveals that the blisters and erosions in pemphigus vulgaris patients are due to loss of cell–cell adhesion just above the basal layer of the epidermis (Fig. 1b). This pathological process is called suprabasilar acantholysis. It is striking that the loss of cell adhesion occurs at this specific level of the epidermis, that the more superficial epidermis maintains its normal architecture and that the basal cells maintain normal attachment to the basement membrane. However, because all of the epidermis above the basal layer is shed, barrier function is extremely compromised. This is manifest clinically as susceptibility to infection and fluid loss, i.e. skin failure.

Before the advent of effective therapy for pemphigus vulgaris, essentially all patients died of skin failure within several years of diagnosis as skin disease became widespread (Lever 1953, 1965).

Like pemphigus vulgaris, pemphigus foliaceus is a disease resulting from loss of keratinocyte cell–cell adhesion. Pemphigus foliaceus patients usually manifest scaly and crusted erosions of the skin, often on the chest and face. Unlike pemphigus vulgaris patients, they do not develop mucous membrane lesions. The primary lesion in pemphigus foliaceus, as in pemphigus vulgaris, is a flaccid blister (Fig. 2a). However, it is more difficult to detect an intact blister in pemphigus foliaceus than in pemphigus vulgaris, because pemphigus foliaceus blisters are even more fragile.

Histopathology of pemphigus foliaceus lesions demonstrates loss of cell–cell adhesion (i.e. acantholysis) in, or just below, the granular layer of the epidermis

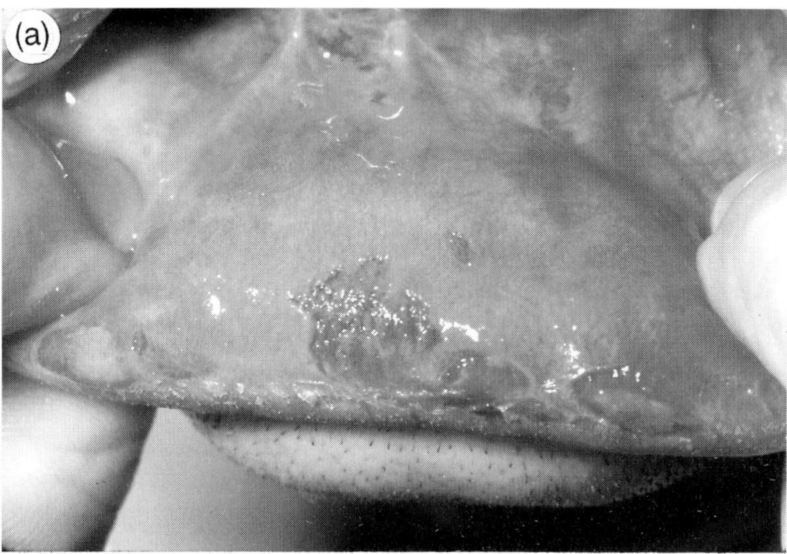

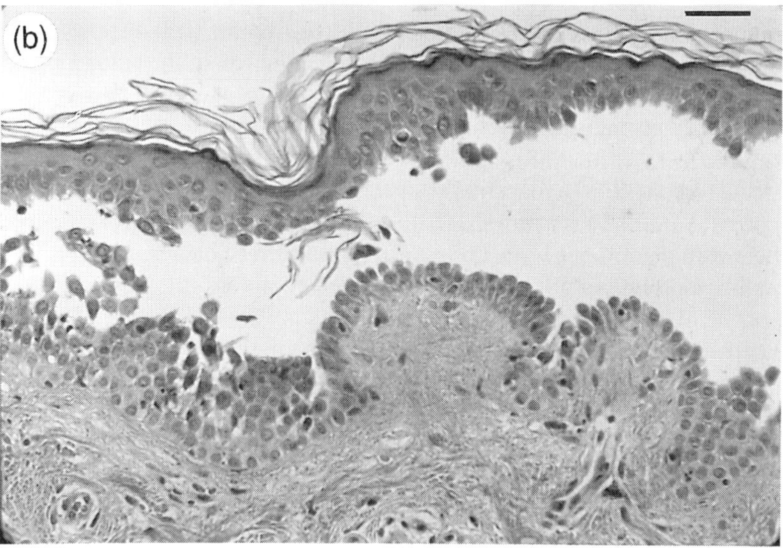

FIG. 1. Pemphigus vulgaris. (a) Characteristic oral erosion on the inner lower lip. (b) Histology of a blister showing suprabasilar acantholysis. (Bar = 45 μm.) (Reprinted from Mueller & Stanley 1990.)

(Fig. 2b). It is striking that the epidermis below this level maintains its normal architecture. These histopathologic findings have two important clinical correlates: (1) because the roof of the blister is so thin, the blister is extremely fragile, accounting for the fact that it is unusual to detect intact blisters; and (2) the barrier function of the skin is not as compromised as in pemphigus vulgaris, because more of the epidermis is preserved when the blister roof is shed. Therefore, the prognosis of pemphigus foliaceus is better than that of pemphigus vulgaris. However, if widespread, pemphigus foliaceus still results in severe morbidity and significant mortality (Lever 1953, 1965, 1979).

This serious prognosis of both types of pemphigus underscores the importance of cell adhesion in the normal structure and function of skin.

Pemphigus autoantibodies cause loss of cell adhesion

In the 1960s it was discovered that pemphigus patients produce autoantibodies against the cell surface of epidermal cells (Beutner & Jordon 1964, Beutner et al 1965). In both types of pemphigus, indirect immunofluorescence indicates the presence of serum antibodies that bind to the cell surface of normal stratified squamous epithelia and direct immunofluorescence indicates that the patients themselves have antibodies bound in their skin (Fig. 3) (Beutner et al 1968, Krasny et al 1987, Rubinstein & Stanley 1987). Thus, pemphigus is associated with the presence of tissue-specific autoantibodies.

Clinical observations suggest that these autoantibodies actually cause disease, i.e. they are capable of causing loss of cell–cell adhesion in the epidermis. There is a general (although not absolute) correlation of autoantibody titre (i.e. the amount of circulating autoantibody) with disease severity (Sams & Jordon 1971, Squiquera et al 1988). Mothers with pemphigus vulgaris can give birth to affected babies due to placental transfer of maternal IgG (Merlob et al 1986). As maternal antibody is catabolized, disease in the neonate resolves. (However, mothers with pemphigus foliaceus do not have affected babies, perhaps because the pathogenic IgG is not efficiently transferred across the placenta [Rocha-Alvarez et al 1992].)

Experimental systems give further evidence of the pathological effects of pemphigus autoantibodies. Pemphigus IgG added to skin organ culture causes acantholysis, even in the absence of complement components and inflammatory cells (Schiltz & Michel 1976, Hashimoto et al 1983). Furthermore, pemphigus vulgaris IgG causes suprabasilar acantholysis, whereas pemphigus foliaceus IgG causes more superficial acantholysis, correlating accurately with the pathology of the respective diseases. Finally, passive transfer of pemphigus IgG to neonatal mice causes epidermal blisters and erosions (Anhalt et al 1982, Roscoe et al 1985). These mice develop IgG fixed to the cell surface of their keratinocytes, as determined by direct immunofluorescence, and lesional skin demonstrates histopathology appropriate to the type of pemphigus IgG transferred.

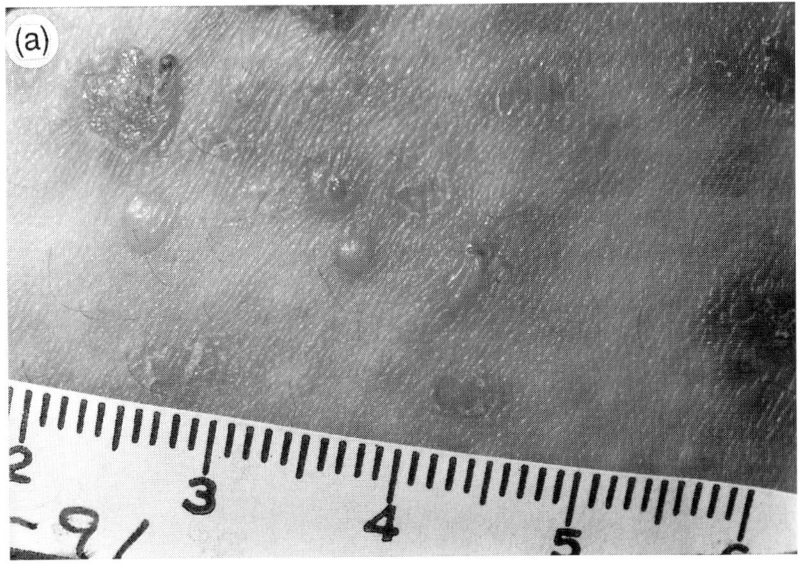

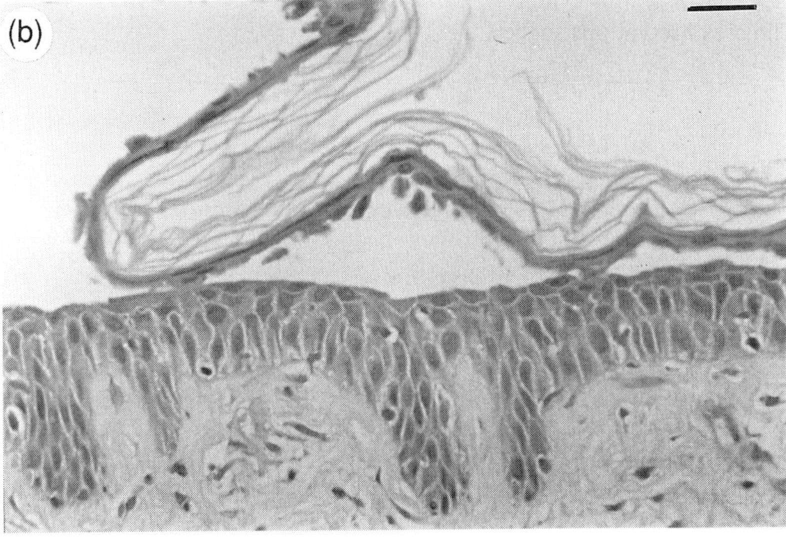

FIG. 2. Pemphigus foliaceus. (a) Flaccid blisters and a crusted erosion on the chest. (b) Histology of a blister showing acantholysis in the granular layer of the epidermis. (Bar = 30 μm.)

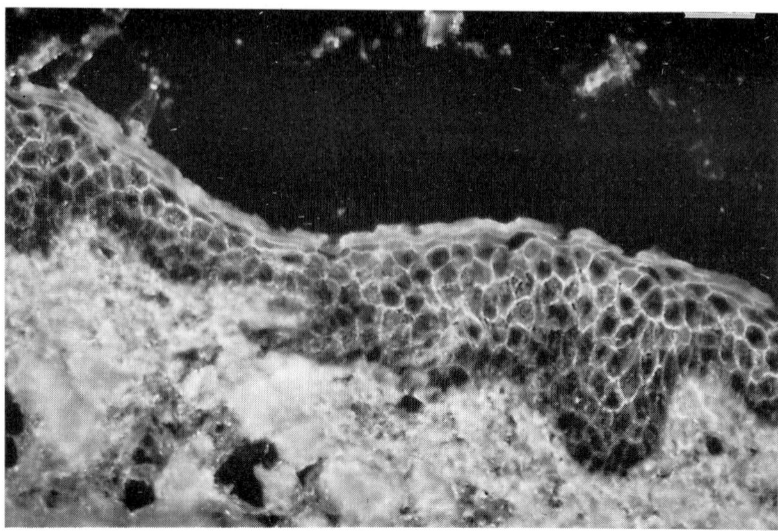

FIG. 3. Direct immunofluorescence for IgG on perilesional skin of a pemphigus foliaceus patient. IgG is present on the cell surface of keratinocytes throughout the epidermis. (Bar = 40 μm.)

Complement-deficient or depleted mice will also develop lesions (Anhalt et al 1986). Finally, even monovalent Fab′ fragments of pemphigus foliaceus IgG are capable of causing disease in neonatal mice, implying that complement activation and/or surface cross-linking are not required to induce disease (Rock et al 1990).

These types of clinical and experimental observations have convincingly demonstrated that pemphigus autoantibodies mediate loss of keratinocyte cell adhesion.

Pemphigus antigens are desmosomal desmogleins

Because pemphigus autoantibodies bind cell surface antigens in normal stratified squamous epithelia and mediate loss of cell adhesion in a tissue-specific fashion, it was postulated that these autoantibodies might define tissue-specific cell adhesion molecules.

At about the same time pemphigus foliaceus antigen was characterized immunochemically as an approximately 160 kDa glycoprotein (Stanley et al 1984), the biochemistry of the constituents of desmosomes was being dissected (Gorbsky & Steinberg 1981, Franke et al 1981). Desmosomes, which are cell adhesion junctions, contained a glycoprotein, called desmoglein, of about the same molecular weight as pemphigus foliaceus antigen. Immunoblotting studies demonstrated that pemphigus foliaceus autoantibodies and antibodies raised against desmoglein bound co-migrating polypeptides in one-dimensional (SDS-PAGE) and two-dimensional (isoelectric focusing/SDS-PAGE) gel electrophoresis (Koulu et al 1984). Immunoprecipitation analysis indicated that most pemphigus foliaceus sera bound calcium-sensitive epitopes on desmoglein and that desmoglein extracted from epidermis was bound in a stoichiometric complex with an 85 kDa molecule, subsequently identified as plakoglobin, a desmosomal plaque protein (Eyre & Stanley 1987, Korman et al 1989).

At the time these observations were made, a paradox was apparent: how could pemphigus foliaceus antigen, which is tissue specific, be the same as desmoglein, which is found in desmosomes of all tissues (Cowin & Garrod 1983)? This paradox, as well as other findings regarding pemphigus foliaceus antigen, were elucidated when cDNA encoding desmoglein was cloned (Koch et al 1990, 1991, 1992, Goodwin et al 1990, Nilles et al 1991, Wheeler et al 1991, Buxton & Magee 1992). There are at least three human desmoglein genes (Buxton & Magee 1992, Buxton et al 1993). The protein dsg-1, encoded by the gene *DSG-1*, is probably limited to stratified squamous epithelia. dsg-2, the product of the *DSG-2* gene, is probably distributed more generally in tissues that express desmosomes. The *DSG-3* gene will be discussed below. By implication, the dsg-1 protein product is thought to be pemphigus foliaceus antigen.

Cloning also revealed that the desmogleins make up a subfamily of the cadherin supergene family. The originally described or classical cadherins (e.g. E-cadherin, N-cadherin) are calcium-dependent, homophilic cell adhesion molecules (Grunwald 1993). Both desmogleins and classical cadherins are transmembrane proteins that share homology in their extracellular domains as well as in one segment of their cytoplasmic domains (Buxton & Magee 1992). In the extracellular domains, both have presumptive calcium-binding sites that probably provide the proper conformation necessary for adhesive function (Ringwald et al 1987, Ozawa et al 1990). In this context, the observation that most pemphigus foliaceus autoantibodies bind calcium-sensitive epitopes implies that these antibodies recognize calcium-dependent conformational determinants, perhaps the same determinants that are functionally important. Although the cytoplasmic domains of classical cadherins and desmogleins share homology in one segment, the desmogleins have a longer cytoplasmic, C-terminal, domain with a repeating amino acid motif termed the (desmoglein) intracellular repeat (Buxton & Magee 1992, Magee & Buxton 1991).

The final link demonstrating that pemphigus foliaceus is a desmosomal component was provided by immunoelectron microscopic studies. As expected, antibodies raised against dsg-1 bind desmosomes, both in the plaque and intercellular region (Steinberg et al 1987, Schmelz et al 1986a). Pemphigus foliaceus autoantibodies also bind to desmosomes, although mostly to the intercellular region (Rappersberger et al 1992). This extracellular localization is consistent with the experimental observations discussed above: (1) most pemphigus foliaceus sera bind calcium-sensitive conformational epitopes which are found in the extracellular domain of dsg-1; and (2) direct immunofluorescence of pemphigus foliaceus patients' skin is positive, implying that the autoantibodies bind to an extracellular domain.

The original immunochemical characterization of pemphigus vulgaris antigen indicated that it was a 130 kDa glycoprotein, disulphide cross-linked to an 85 kDa protein (Stanley et al 1984, Eyre & Stanley 1988). The discovery that the 85 kDa was plakoglobin, a desmosomal plaque protein, led to the hypothesis that pemphigus vulgaris antigen is also an adhesion junction molecule (Korman et al 1989). This hypothesis was further supported when cDNA cloning of pemphigus vulgaris antigen indicated that it, like pemphigus foliaceus antigen, was in the desmoglein subfamily of cadherins (Amagai et al 1991). Pemphigus vulgaris antigen is, therefore, now also termed dsg-3. Rabbit antibodies raised against fusion proteins containing extracellular subdomains of pemphigus vulgaris antigen allowed immunoelectron microscopic localization of pemphigus vulgaris antigen to desmosomes (Fig. 4) (Karpati et al 1993).

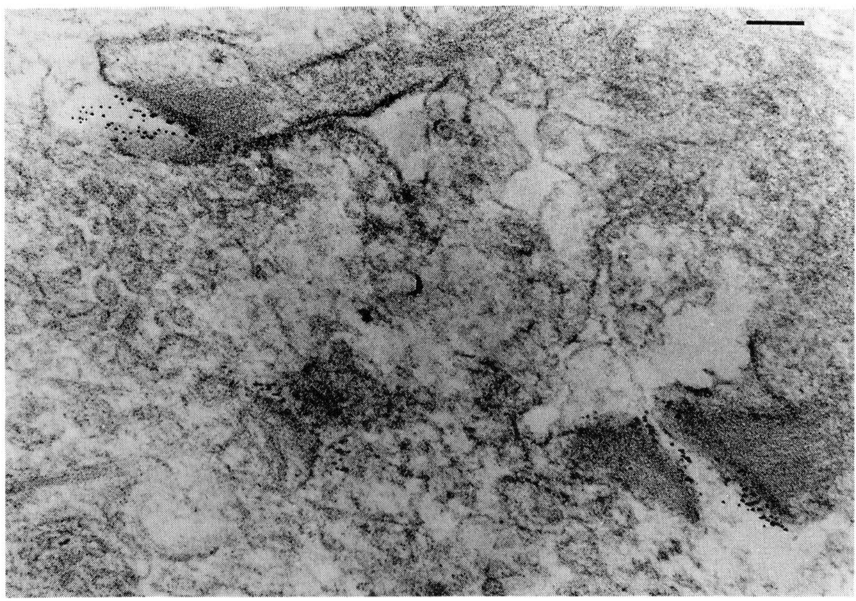

FIG. 4. Pemphigus vulgaris antigen is in the desmosomes of cultured keratinocytes. Rabbit antibodies raised to the extracellular domain of pemphigus vulgaris antigen bind the extracellular face of desmosomes slightly separated by a brief trypsin/calcium incubation, as detected by immunogold electron microscopy. (Bar = 0.1 μm.) (Reprinted from Karpati et al 1993.)

Cloning of the cDNA for pemphigus vulgaris antigen allowed us to carry out experiments to determine against which extracellular subdomains patient antibodies are directed and if these antibodies could cause loss of cell–cell adhesion in the neonatal mouse model of disease (Amagai et al 1992). These studies showed that pemphigus vulgaris patient antibodies are directed against various epitopes in the pemphigus vulgaris antigen extracellular domain and that antibodies affinity-purified against the more N-terminal subdomains (thought to be important in cadherin adhesive ability) are capable of causing loss of epidermal cell adhesion in neonatal mice (Fig. 5) (Amagai et al 1992). Finally, these affinity-purified antibodies were localized ultrastructurally to separating desmosomes in these mice (Karpati et al 1993).

In summary, both pemphigus foliaceus and pemphigus vulgaris are autoimmune diseases characterized by loss of epidermal cell–cell adhesion. In both diseases, patients have pathogenic autoantibodies directed against desmogleins found in desmosomes.

Major differences between desmogleins and classical cadherins

Many observations, discussed above, suggest the hypothesis that desmogleins are calcium-dependent cell adhesion molecules, like classical cadherins. (1) Desmogleins are homologous to classical cadherins in segments of both the

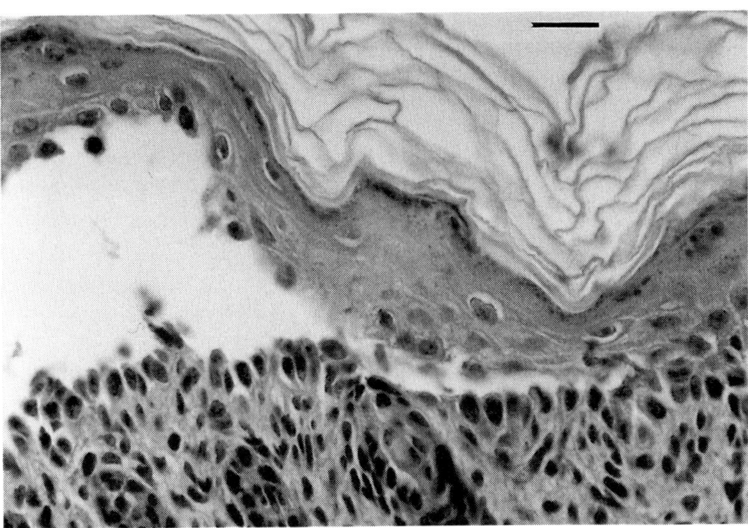

FIG. 5. Pemphigus vulgaris patient's antibodies affinity purified on the extracellular N-terminal subdomains of pemphigus vulgaris antigen and injected into a neonatal mouse cause suprabasilar acantholysis. Hematoxylin and eosin stained section of skin. (Bar = 25 μm.)

extracellular and cytoplasmic domains. (2) Desmogleins share presumptive calcium-binding domains with cadherins (Amagai et al 1991, Nilles et al 1991). (3) Desmogleins and cadherins share a homologous triplet of amino acids in a conserved position in the first extracellular domain that is thought to be the centre of homophilic binding (Buxton & Magee 1992). (4) Desmogleins are found in a cell adhesion junction, the desmosome. (5) In pemphigus foliaceus and pemphigus vulgaris, autoantibodies against desmogleins cause loss of cell adhesion.

However, there are significant biochemical differences between desmogleins and classical cadherins. Desmogleins have a longer cytoplasmic tail than the cadherins. Whereas the cytoplasmic tail of cadherins binds α- and β-catenins, the cytoplasmic tail of desmogleins binds only plakoglobin (Peifer et al 1992). A characteristic property of the extracellular domain of classical cadherins is that they are resistant to degradation by trypsin in calcium-containing medium (Takeichi 1990, Hyafil et al 1981, Yoshida & Takeichi 1982, Damsky et al 1983). Pemphigus vulgaris antigen and perhaps other desmogleins do not share this property—the extracellular domain is degraded by trypsin/calcium (Plott et al 1994). These biochemical differences suggest that perhaps desmogleins and classical cadherins also have distinct functions.

That this is indeed the case is shown by studies in which the extracellular domain of E-cadherin was substituted, in a chimeric molecule, with the extracellular domain of pemphigus vulgaris (Amagai et al 1994). Whereas the full length authentic E-cadherin, expressed in transfected L cells, mediated strong cell–cell adhesion, the chimeric molecule did not. Therefore, the extracellular domains of these molecules are not simply interchangeable in functional studies. This type of result suggests that either desmogleins may not function in homophilic adhesion or that, if they do, the conditions for function (e.g. associated molecules, cellular organization) are quite different from those necessary for function of classical cadherins.

Future directions

There are many questions that remain regarding diseases of cell–cell adhesion in epidermis.

Certain clinical observations are not well understood on a pathophysiological level. Pemphigus foliaceus patients do not develop mucous membrane lesions. This is difficult to understand because stratified squamous epithelia contain desmosomes and desmoglein (Schmelz et al 1986b). In fact, autoantibodies from the sera of pemphigus foliaceus patients bind non-epidermal stratified squamous epithelia (Rubinstein & Stanley 1987) and these patients have IgG bound in their own mucous membranes (Diaz et al 1988). Another clinical paradox is the fact that pemphigus foliaceus and pemphigus vulgaris patients develop loss of cell adhesion at different levels of the epidermis when their antibodies show the same distribution of binding within the epidermis (Wood & Beutner 1977).

Future studies will examine other, non-autoimmune, diseases in which epidermal cells lose their cell–cell adhesion. Two of these diseases, Hailey-Hailey disease and Darier's disease, are inherited in an autosomal dominant pattern. Among the candidate genes involved in these diseases will certainly be those encoding cell adhesion molecules specific for stratified squamous epithelia (to which these diseases are localized).

Certain fundamental cell biological questions regarding desmogleins also remain. What is their function? Is it adhesion? If so, is this adhesion heterophilic or homophilic? Does it depend on calcium? Does it require other desmosomal molecules? How is the adhesion in desmosomes normally down-regulated, for instance, when cells must migrate upward with differentiation or laterally in wound healing? Is it possible that desmogleins mediate signal transduction?

Finally, if we can understand the normal function of desmogleins we might be in a better position to understand how autoantibodies in pemphigus cause loss of cell adhesion. Do they interfere directly with function? Do they somehow result in signal transduction inside the cell to cause protease release which has been suggested to be important in pemphigus blister formation (Morioka et al 1987)?

Whatever answers we get to these questions, they are bound to be interesting both from the point of view of epidermal cell adhesion and pathophysiology of disease.

References

Amagai M, Klaus-Kovtun V, Stanley JR 1991 Autoantibodies against a novel epithelial cadherin in pemphigus vulgaris, a disease of cell adhesion. Cell 67:869–877

Amagai M, Karpati S, Prussick R, Klaus-Kovtun V, Stanley JR 1992 Autoantibodies against the amino-terminal cadherin-like binding domain of pemphigus vulgaris antigen are pathogenic. J Clin Invest 90:919–926

Amagai M, Karpati S, Klaus-Kovtun V, Udey MC, Stanley JR 1994 The extracellular domain of pemphigus vulgaris antigen (desmoglein 3) mediates weak homophilic adhesion. J Invest Dermatol 102:402–408

Anhalt GJ, Labib RS, Voorhees JJ, Beals TF, Diaz LA 1982 Induction of pemphigus in neonatal mice by passive transfer of IgG from patients with the disease. N Engl J Med 306:1189–1196

Anhalt GJ, Till GO, Diaz LA, Labib RS, Patel HP, Eaglstein NF 1986 Defining the role of complement in experimental pemphigus vulgaris in mice. J Immunol 137:2835–2840

Beutner EH, Jordon RE 1964 Demonstration of skin antibodies in sera of pemphigus vulgaris patients by indirect immunofluorescent staining. Proc Soc Exp Bio Med 117:505–510

Beutner EH, Lever WF, Witebsky E, Jordon RE, Chertock B 1965 Autoantibodies in pemphigus vulgaris. JAMA (J Am Med Assoc) 192:682–688

Beutner EH, Jordon RE, Chorzelski TP 1968 The immunopathology of pemphigus and bullous pemphigoid. J Invest Dermatol 51:63–80

Buxton RS, Magee AI 1992 Structure and interactions of desmosomal and other cadherins. Semin Cell Biol 3:157–167

Buxton RS, Cowin P, Franke WW et al 1993 Nomenclature of the desmosomal cadherins. J Cell Biol 121:481–483

Cowin P, Garrod DR 1983 Antibodies to epithelial desmosomes show wide tissue and species cross-reactivity. Nature 302:148–150
Damsky CH, Richa J, Solter D, Knudsen K, Buck CA 1983 Identification and purification of a cell surface glycoprotein mediating intercellular adhesion in embryonic and adult tissue. Cell 34:455–466
Diaz LA, Sampaio SAP, Rivitti EA et al 1988 Endemic pemphigus foliaceus (fogo selvagem). I. Clinical features and immunopathology. J Am Acad Dermatol 20:657–669
Eyre RW, Stanley JR 1987 Human autoantibodies against a desmosomal protein complex with a calcium-sensitive epitope are characteristic of pemphigus foliaceus patients. J Exp Med 165:1719–1724
Eyre RW, Stanley JR 1988 Identification of pemphigus vulgaris antigen extracted from normal human epidermis and comparison with pemphigus foliaceus antigen. J Clin Invest 81:807–812
Franke WW, Schmid E, Grund C et al 1981 Antibodies to high molecular weight polypeptides of desmosomes: specific localization of a class of juctional proteins in cells and tissues. Differentiation 20:217–241
Goodwin L, Hill JE, Raynor K, Raszi L, Manabe M, Cowin P 1990 Desmoglein shows extensive homology to the cadherin family of cell adhesion molecules. Biochem Biophys Res Commun 173:1224–1230
Gorbsky G, Steinberg MS 1981 Isolation of the intercellular glycoproteins of desmosomes. J Cell Biol 90:243–248
Grunwald GB 1993 The structural and functional analysis of cadherin calcium-dependent cell adhesion molecules. Curr Opin Cell Biol 5:797–805
Hashimoto K, Shafran KM, Webber PS, Lazarus GS, Singer KH 1983 Anti-cell surface pemphigus autoantibody stimulates plasminogen activator activity of human epidermal cells. J Exp Med 157:259–272
Hyafil F, Babinet C, Jacob F, 1981 Cell–cell interactions in early embryogenesis: a molecular approach to the role of calcium. Cell 26:447–454
Karpati S, Amagai M, Prussick R, Cehrs K, Stanley JR 1993 Pemphigus vulgaris antigen, a desmoglein type of cadherin, is localized within keratinocyte desmosomes. J Cell Biol 122:409–415
Koch PJ, Walsh MJ, Schmelz M, Goldschmidt MD, Zimbelmann R, Franke WW 1990 Identification of desmoglein, a constitutive desmosomal glycoprotein, as a member of the cadherin family of cell adhesion molecules. Eur J Cell Biol 53:1–12
Koch PJ, Goldschmidt MD, Walsh MJ, Zimbelmann R, Franke WW 1991 Complete amino acid sequence of the epidermal desmoglein precursor polypeptide and identification of a second type of desmoglein gene. Eur J Cell Biol 55:200–208
Koch PJ, Goldschmidt MD, Zimbelmann R, Troyanovsky R, Franke WW 1992 Complexity and expression patterns of the desmosomal cadherins. Proc Natl Acad Sci USA 89:353–357
Korman NJ, Eyre RW, Klaus-Kovtun V, Stanley JR 1989 Demonstration of an adhering-junction molecule (plakoglobin) in the autoantigens of pemphigus foliaceus and pemphigus vulgaris. N Engl J Med 321:631–635
Koulu L, Kusumi A, Steinberg MS, Klaus-Kovtun V, Stanley JR 1984 Human autoantibodies against a desmosomal core protein in pemphigus foliaceus. J Exp Med 160:1509–1518
Krasny SA, Beutner EH, Chorzelski TP 1987 Specificity and sensitivity of indirect and direct immunofluorescent findings in the diagnosis of pemphigus. In: Beutner EH, Chorzelski TP, Kumar V (eds) Immunopathology of the skin. Wiley, New York, p 207–247
Lever WF 1953 Pemphigus. Medicine 32:1–12
Lever WF 1965 Pemphigus and pemphigoid. Charles C Thomas, Springfield, IL

Lever WF 1979 Pemphigus and pemphigoid. J Am Acad Dermatol 1:2–31
Magee AI, Buxton RS 1991 Transmembrane molecular assemblies regulated by the greater cadherin family. Curr Opin Cell Biol 3:854–861
Merlob P, Metzker A, Hazaz BAC, Rogovin H, Reisner SH 1986 Neonatal pemphigus vulgaris. Pediatrics 78:1102–1105
Morioka S, Lazarus GS, Jensen PJ 1987 Involvement of urokinase-type plasminogen activator in acantholysis induced by pemphigus IgG. J Invest Dermatol 89:474–477
Mueller S, Stanley JR 1990 Pemphigus: pemphigus vulgaris and pemphigus foliaceus. In: Wojnarowska F, Briggaman RA (eds) Management of blistering diseases. Chapman & Hall, London, p 43–61
Nilles LA, Parry DAD, Powers EE, Angst BD, Wagner RM, Green KJ 1991 Structural analysis and expression of human desmoglein: a cadherin-like component of the desmosome. J Cell Sci 99:809–821
Ozawa M, Engel J, Kemler R 1990 Single amino acid substitutions in one Ca^{2+} binding site of uvomorulin abolish the adhesive function. Cell 63:1033–1038
Peifer M, McCrea PD, Green KJ, Wieschaus E, Gumbiner BM 1992 The vertebrate adhesive junction proteins β-catenin and plakoglobin and the drosophila segment polarity gene *armadillo* form a multigene family with similar properties. J Cell Biol 118:681–691
Plott RT, Amagai M, Udey MC, Stanley JR 1994 Pemphigus vulgaris antigen lacks biochemical properties characteristic of classical cadherins. J Invest Dermatol 103:168–172
Rappersberger K, Roos N, Stanley JR 1992 Immunomorphological and biochemical identification of the pemphigus foliaceus autoantigen within desmosomes. J Invest Dermatol 99:323–330
Ringwald M, Schuh R, Vestweber D et al 1987 The structure of cell adhesion molecule uvomorulin. Insights into the molecular mechanism of Ca^{2+}-dependent cell adhesion. EMBO (Eur Mol Biol Organ) J 6:3647–3653
Rocha-Alvarez R, Friedman H, Campbell IT, Souza-Aguiar L, Martins-Castro R, Diaz LA 1992 Pregnant women with endemic pemphigus foliaceus (Fogo Selvagem) give birth to disease-free babies. J Invest Dermatol 99:78–82
Rock B, Labib RS, Diaz LA 1990 Monovalent Fab′ immunoglobulin fragments from endemic pemphigus foliaceus autoantibodies reproduce the human disease in neonatal Balb/c mice. J Clin Invest 85:296–299
Roscoe JT, Diaz L, Sampaio SA et al 1985 Brazilian pemphigus foliaceus autoantibodies are pathogenic to BALB/c mice by passive transfer. J Invest Dermatol 85:538–541
Rubinstein N, Stanley JR 1987 Pemphigus foliaceus antibodies and a monoclonal antibody to desmoglein I demonstrate stratified squamous epithelial-specific epitopes of desmosomes. Am J Dermatopathol 9:510–514
Sams WM Jr, Jordon RE 1971 Correlation of pemphigoid and pemphigus antibody titres with activity of disease. Br J Dermatol 84:7–13
Schiltz JR, Michel B 1976 Production of epidermal acantholysis in normal human skin in vitro by the IgG fraction from pemphigus serum. J Invest Dermatol 67:254–260
Schmelz M, Duden R, Cowin P, Franke WW 1986a A constitutive transmembrane glycoprotein of Mr 165,000 (desmoglein) in epidermal and non-epidermal desmosomes. II. Immunolocalization and microinjection studies. Eur J Cell Biol 42:184–199
Schmelz M, Duden R, Cowin P, Franke WW 1986b A constitutive transmembrane glycoprotein of Mr 165,000 (desmoglein) in epidermal and non-epidermal desmosomes. I. Biochemical identification of the polypeptide. Eur J Cell Biol 42:177–183
Squiquera HL, Diaz LA, Sampaio SAP et al 1988 Serologic abnormalities in patients with endemic pemphigus foliaceus (fogo selvagem), their relatives, and normal donors from endemic and non-endemic areas of Brazil. J Invest Dermatol 91:189–191

Stanley JR 1990 Pemphigus: skin failure mediated by autoantibodies. JAMA (J Am Med Assoc) 264:1714–1717
Stanley 1992 Pemphigus. In: Fitzpatrick TB, Eisen AZ, Wolff K, Freedberg IM, Austen KF (eds) Dermatology in general medicine. McGraw-Hill, New York, p 606–615
Stanley JR, Koulu L, Thivolet C 1984 Distinction between epidermal antigens binding pemphigus vulgaris and pemphigus foliaceus autoantibodies. J Clin Invest 74:313–320
Steinberg MS, Shida H, Giudice GJ, Shida M, Patel NH, Blaschuk OW 1987 On the molecular organization, diversity and functions of desmosomal proteins. In: Junctional complexes of epithelial cells. Wiley, Chichester (Ciba Found Symp 125) p 3–25
Takeichi M 1990 Cadherins: a molecular family important in selective cell–cell adhesion. Annu Rev Biochem 59:237–252
Wheeler GN, Parker AE, Thomas CL et al 1991 Desmosomal glycoprotein DGI, a component of intercellular desmosome junctions, is related to the cadherin family of cell adhesion molecules. Proc Natl Acad Sci USA 88:4796–4800
Wood GW, Beutner EH 1977 Blocking-immunofluorescence studies on the specificity of pemphigus autoantibodies. Clin Immunol Immunopathol 7:168–175
Yoshida C, Takeichi M 1982 Teratocarcinoma cell adhesion: identification of a cell-surface protein involved in calcium-dependent cell aggregation. Cell 28:217–224

DISCUSSION

Wagner: Can you co-express plakoglobin and desmogleins?

Stanley: This is an important step. Kathy Green has tried this (unpublished results). Plakoglobin is unstable when expressed alone in L cells, which are mouse fibroblasts. dsg-1, however, stabilizes the plakoglobin expression.

Hynes: Is plakoglobin the same as γ-catenin?

Stanley: This is controversial. Plakoglobin was first thought to be γ-catenin. Recent studies, however, using 2D gel electrophoresis, have shown that plakoglobin and γ-catenin may be biochemically distinct (Piepenhagen & Nelson 1993).

Birchmeier: We think that plakoglobin is in fact γ-catenin. We have also shown that plakoglobin and β-catenin link E-cadherin to α-catenin and thereby to the cytoskeleton.

Stanley: We have done similar experiments in keratinocytes using immunoprecipitation. E-cadherin binds to β-catenin and to plakoglobin.

Pober: Is there a heterophilic ligand for the desmosomal cadherins?

Stanley: The assumption is that binding is homophilic because it is so for classical cadherins. However, the case for desmosomal cadherins is not known.

Pober: It would be a straightforward experiment to do. You could transfect desmosomal cadherins into L cells and test for homophilic versus heterophilic binding.

Stanley: I agree, we just haven't done that yet with the full length pemphigus vulgaris antigen.

Garrod: We have transfected non-adhesive L cells with bovine type 1 desmocollins, obtaining cell lines that express either the longer 'a' form or the shorter 'b' form. Aggregation assays show that both 'a' and 'b' forms mediate homophilic Ca^{2+}-dependent adhesion which is weaker than that produced by E-cadherin. Cytoskeletal attachment does not seem to be essential for this desmocollin-mediated aggregation (M. A. J. Chidgey & D. R. Garrod, unpublished results).

Humphries: Is there a correlation between the tissue distribution of the antigen and the site of blistering?

Stanley: The pemphigus vulgaris antigen is found only in stratified squamous epithelia, the tissue where the disease occurs. However, the antigen seems to be distributed throughout the epidermis, whereas blistering only occurs in a suprabasal location.

Birchmeier: dsg-3 is a minor antigen; dsg-2 is the major desmoglein, occurring in tissues throughout the body. Why are there no autoantibodies made against dsg-2?

Stanley: There is a genetic restriction on the specificity of antibodies that any individual can make. The antibodies in pemphigus are all tissue specific, as are dsg-3 and dsg-1. On the other hand, dsg-2 occurs in all desmosomes. There is no known autoimmune disease in which the autoantibodies bind all desmosome-containing tissues.

Ruggeri: Are antibodies from different patients of the same idiotype?

Stanley: We have not looked for cross-reactivity.

Sonnenberg: You said that the desmosomes are very tight: is it possible for antibodies to penetrate them?

Stanley: Yes; although we have to open the desmosomes to allow the large gold particles to enter for immunostaining, antibodies are smaller and able to penetrate.

Garrod: In vivo, desmosomes are not static structures: they must disassemble and reassemble in order to allow cells to change position. This may also permit the penetration of antibodies.

Riethmüller: If the antibody has general access to the skin, why are there areas with no blisters in many cases?

Stanley: We have taken biopsies of the areas of normal skin in these patients. There are antibodies there. However, this normal skin is often fragile; when you rub the skin it may come off. On the other hand, the skin may not be at all fragile and still have *in vivo* bound antibodies. Therefore, more than just antibody binding is required for disease. This fact is also illustrated by the observation that if you treat these pemphigus vulgaris patients, who have spontaneous blisters, with corticosteroids, there is improvement within a few days. This is too early for the antibody to have been cleared, so the corticosteroids must be interfering with a more local mechanism.

Hynes: Do the blisters appear at random over the body or is there a specific pattern of distribution?

Stanley: In pemphigus vulgaris they appear predominantly in certain places, such as the mouth and the scalp. Pemphigus foliaceus occurs on the upper chest and face. There is a rough correlation between the amount of antigen in various places and the appearance of disease (Sison-Fonacier & Bystryn 1986).

Pals: Are normal people tolerant to this antigen?

Stanley: Presumably. We have not tried to cause active pemphigus in animals. Rabbits can be immunized with the pemphigus vulgaris antigen and they develop antibodies but not the disease—the antibodies bind to hidden epitopes.

Sonnenberg: Do any of the antibodies block function?

Stanley: No.

Ruggeri: Do the antibodies from patients bind to normal desmosomes in people?

Stanley: Yes, as determined by indirect immunofluorescence.

Verrando: There are many molecules involved in cellular adhesion. Why don't these take over the role of the pemphigus vulgaris antigen in pemphigus vulgaris patients?

Stanley: I agree. Nature is clever and there is a lot of redundancy, in the sense of overlapping functions. In the epidermis one finds E- and P-cadherin, the desmogleins and the desmocollins. All of these might subserve adhesion functions. However, the pathology induced by pemphigus vulgaris autoantibodies is probably more than simply the antibody interfering with the function of the pemphigus vulgaris antigen. For example, antibody binding might induce protease release from the keratinocytes and these proteases might destroy various adhesive molecules.

Hogg: Is pemphigus confined to skin? Desmogleins are also found internally.

Stanley: dsg-1 and dsg-3 are limited to squamous stratified epithelia. All other desmosomes contain dsg-2.

Labow: Do pemphigus patients share a common haplotype?

Stanley: This has been looked at in a lot of Israeli patients. Pemphigus vulgaris in ethnic groups does turn out to be related to HLA haplotype (Sinha et al 1988, Szafer et al 1987, Scharf et al 1989, Ahmed et al 1990).

Garrod: We have done immunoblots on desmosomes from bovine nasal epidermis with sera from patients with pemphigus vulgaris, pemphigus foliaceous and endemic Brazilian pemphigus foliaceous. We find that a proportion of these sera possess reactivity to desmocollins: 20% in pemphigus vulgaris, 11% in pemphigus foliaceous and 40% in endemic Brazilian pemphigus foliaceous. We do not know whether these desmocollin autoantibodies are pathogenic but they suggest that the pathogenesis of pemphigus may be more complex than previously believed.

Sera from patients with pemphigus have not been found to react with simple epithelial cells. An exception is the MDCK (canine kidney) cell line which shows bright staining around the periphery with pemphigus vulgaris sera. The antigen recognized by these sera in the MDCK cells is a 130 kDa polypeptide that is also recognized by a monoclonal antibody that reacts with dsg-1 and dsg-2 in epidermis. Pemphigus vulgaris sera do not affect adhesion between MDCK cells. This may represent the only case of pemphigus vulgaris serum reactivity with simple epithelial cells. The MDCK cells are fairly odd cells; they have been culture-adapted for many years. However, they express a protein closely related to pemphigus vulgaris antigen.

References

Ahmed AR, Yunis EJ, Khatri K et al 1990 Major histocompatibility complex haplotype studies in Ashkenazi Jewish patients with pemphigus vulgaris. Proc Natl Acad Sci USA 87:7685–7662

Piepenhagen PA, Nelson WJ 1993 Defining E-cadherin-associated protein complexes in epithelial cells: plakogobin, β-catenin and γ-catenin are distinct components. J Cell Sci 104:751–762

Scharf SJ, Freidmann A, Steinman L, Brautbar C, Erlich HA 1989 Specific *HLA-DQB* and *HLA-DRB1* alleles confer susceptibility to pemphigus vulgaris. Proc Natl Acad Sci USA 86:6215–6219

Sinha AA, Brautbar C, Szafer F et al 1988 A newly characterized HLA DQ_{β} allele associated with pemphigus vulgaris. Science 239:1026–1029

Sison-Fonacier L, Bystryn JC 1986 Regional variations in antigenic properties of skin. A possible cause for disease-specific distribution of skin lesions. J Exp Med 164:2125–2130

Szafer F, Brautbar C, Tzfoni E et al 1987 Detection of disease-specific restriction fragment length polymorphisms in pemphigus vulgaris linked to the DQw1 and DQw3 alleles of the HLA-D region. Proc Natl Acad Sci USA 84:6542–6545

E-cadherin as an invasion suppressor

Walter Birchmeier, Jörg Hülsken and Jürgen Behrens

Max-Delbrück-Centre for Molecular Medicine, Robert-Rössle-Straße 10, 13125 Berlin, Germany

Abstract. The loss of epithelial differentiation in carcinomas, which is accompanied by increased mobility and invasiveness of the tumour cells, is often a consequence of reduced intercellular adhesion. Recent reports have indicated that the primary cause for the 'scattering' of the cells in invasive carcinomas is a disturbance of the integrity of intercellular junctions often involving the cell adhesion molecule E-cadherin. It has also been suggested that during invasion, carcinoma cells convert to a sort of mesenchymal stage, as do normal epithelial cells during development. Permanent and transient molecular mechanisms lead to the impairment of junction integrity of epithelial cells and thus to the progression of carcinomas towards a more invasive state.

1995 Cell adhesion and human disease. Wiley, Chichester (Ciba Foundation Symposium 189) p 124–141

It has recently become evident that the proper formation of intercellular junctions is critical for the maintenance of epithelial differentiation; destabilization of junctions allows invasiveness of epithelial cells and the progression of carcinomas (see Tsukita et al 1993, Birchmeier & Behrens 1994, for reviews). Accordingly, various structural components of intercellular junctions have been found to be related to products of tumour suppressor genes. For instance, the cell adhesion molecule E-cadherin suppresses invasiveness of carcinoma cells and the E-cadherin gene is mutated in about 50% of the diffuse-type gastric carcinomas (Behrens et al 1989, Frixen et al 1991, Vleminckx et al 1991, Becker et al 1994, Risinger et al 1994, Oda et al 1994). The tumour suppressor gene product APC binds to β-catenin, which is cytoplasmically associated with E-cadherin (Rubinfeld et al 1993, Su et al 1993, Ozawa et al 1989, Nagafuchi & Takeichi 1988). The tight-junction-associated protein ZO-1 is related to the *dlg* tumour suppressor gene product of *Drosophila*, and the neurofibromatosis-2 tumour suppressor gene product merlin/schwannomin to the ezrin/radixin/moesin family of junctional proteins (Woods et al 1991, Rouleau et al 1993, Trofatter et al 1993). Conversely, products of oncogenes such as *src*, *ras*, *fos*, or *met* have been shown to destabilize intercellular junctions (Behrens et al 1989, 1993, Weidner et al 1990, Hamaguchi et al 1993, Shibamoto et al 1994, Reichmann et al 1992). Src (the EGF receptor) and Met phosphorylate β-catenin on tyrosine residues (Behrens et al 1993, Hamaguchi et al 1993, Shibamoto et al 1994).

Invasive carcinoma cells exhibit reduced intercellular adhesion: role of adherens junctions

Pathologists have observed for some time that invasive carcinomas show reduced epithelial differentiation; this is particularly prominent at invasion fronts where the cells break into surrounding stromal tissue (Gabbert et al 1985). The invading cells can lose their epithelial appearance and become amorphic, spindle-shaped, or fibroblastoid. The loosening of intercellular adhesiveness in carcinoma cells is due to a functional disturbance of cell–cell contacts (see Birchmeier & Behrens 1994 for a review). Apparently, the malignant cells down-modulate intercellular adhesiveness as a prerequisite for invasive behaviour. That down-modulation of intercellular adhesiveness is crucial for carcinomas to progress was first proposed 50 years ago (Coman 1944).

The various types of adhesion deficiencies in invasive carcinomas may be understood better if one takes into account the overall structure of the cell–cell adherens junctions of epithelial cells (Fig. 1A). Adherens junctions are specialized regions of plasma membranes where transmembranous E-cadherin molecules on opposing cells make contact with each other through interaction between one or more of the extracellular cadherin repeats (Edelman et al 1988, Kemler et al 1989, Takeichi 1991, Birchmeier & Behrens 1994; for a discussion of the related desmosomes, see Buxton & Magee 1992). The cytoplasmic domain of E-cadherin is associated with a group of proteins, named catenins, which make contact with the actin microfilament network through unknown linkage proteins. The cDNAs of two catenins have recently been characterized: α-catenin shows structural similarity to the junction-associated protein vinculin (Herrenknecht et al 1991, Nagafuchi et al 1991) and β-catenin is homologous to the product of the *Drosophila* segment polarity gene *armadillo* (McCrea et al 1991, Butz et al 1992). γ-catenin is related or identical to plakoglobin (Knudsen et al 1992; but see also Piepenhagen & Nelson 1993).

It is obvious that any significant change in expression or structure of one of the essential components of the adherens junctions could lead to junctional disassembly and, consequently, to more mobile invasive carcinoma cells. Molecular aberrations which result in such defects *in vitro* and *in vivo* have been studied recently: expression of E-cadherin without a cytoplasmic domain in fibroblasts did not result in functional cell–cell adhesion, in contrast to the expression of intact E-cadherin. Apparently, truncated E-cadherin is unable to interact with the catenins and, therefore, proper junctions cannot be formed (Nagafuchi & Takeichi 1988, Ozawa et al 1989); cadherins without the entire extracellular domain acted in a dominant-negative fashion, i.e. disturbed

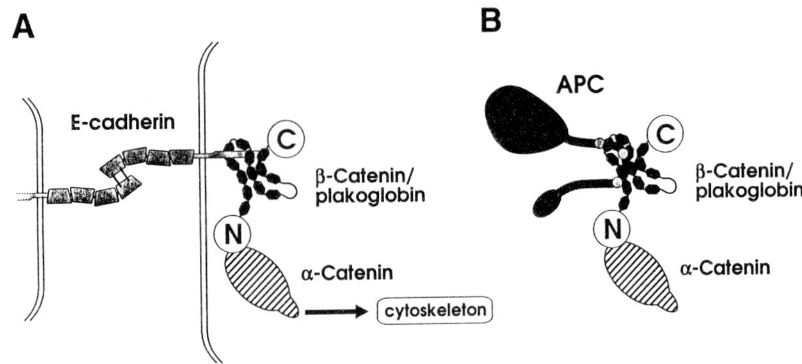

FIG. 1. Schematic representation of (A) the E-cadherin and (B) APC complexes involving β-catenin or plakoglobin. β-catenin/plakoglobin interact with the cytoplasmic region of the cell adhesion molecule E-cadherin via their internal *armadillo*-like repeats (drawn as links of a chain). α-catenin, which mediates the connection to the cytoskeleton, binds to the N-terminal domain of β-catenin. No function has yet been assigned to the C-terminal domain of β-catenin. β-catenin/plakoglobin also interact with APC through the *armadillo*-like repeats. The three β-catenin-binding repeats of APC are drawn. We also indicate that the repeat units of β-catenin located more N- or C-terminally interact with APC or E-cadherin, respectively.

epithelial cell adhesion. It is likely that these truncated molecules collect catenins, which are thus not available to form proper complexes with intact E-cadherin (Kintner 1992, Fujimori & Takeichi 1993). A point mutation generated in the Ca^{2+}-binding region of the first extracellular repeat of E-cadherin resulted in non-functional cell adhesion molecules (Ozawa et al 1990). Mutations of the E-cadherin gene have recently been discovered in diffuse-type human gastric carcinomas (Becker et al 1994). In 50% of cases, gene mutations lead to slightly shortened mRNAs in which the information for exon 8 or 9 is deleted. The functional significance of such exon skipping (the Ca^{2+}-binding region of E-cadherin is affected) remains to be determined. Mutations of the E-cadherin gene have recently also been identified in human ovarian carcinomas (Risinger et al 1994) and in two non-adhesive cell lines from gastric carcinomas (Oda et al 1994). The discovery of E-cadherin gene mutations in freshly isolated human tumours now permits classification of E-cadherin as a true tumour (invasion) suppressor gene (cf. Behrens et al 1989, Frixen et al 1991, Vleminckx et al 1991). The adhesive properties of PC9 lung cancer cells, which harbour a null mutation in the α-catenin gene, could be restored by transfection with α-catenin cDNA (Hirano et al 1992, Shimoyama et al 1992, Oda et al 1993). Finally, transformation of epithelial cells with a temperature-sensitive *src* gene resulted in a reversible loss of differentiation and in unchanged E-cadherin expression, but tyrosine residues of β-catenin were phosphorylated at the permissive temperature. It has been hypothesized that tyrosine phosphorylation of β-catenin results in the disturbance of the adhesion complexes and leads to a non-adhesive state (Behrens et al 1993, Hamaguchi et al 1993, Shibamoto et al 1994).

E-cadherin expression in various tumours

In the last few years, various laboratories have become aware of the role of E-cadherin in formation of epithelial cell junctions and of the putative function of this protein as an invasion suppressor of carcinomas, and they have begun to examine E-cadherin expression in various human and animal carcinomas. A correlation between down-regulation of E-cadherin-mediated cell adhesion and invasion potential of many carcinomas has emerged from these studies.

We recently investigated the expression of E-cadherin in tissue sections of 32 squamous cell carcinomas of the head and neck (Schipper et al 1991). We found that E-cadherin expression, as detected by immunofluorescence and *in situ* hybridization, is inversely correlated both with the loss of differentiation of the tumours and with lymph node metastasis. The well-differentiated squamous cell carcinomas expressed E-cadherin, often as strongly as the normal stratified epithelium, and the moderately differentiated squamous cell carcinomas expressed intermediate amounts of E-cadherin or were heterogeneous, whereas the poorly differentiated squamous cell carcinomas were all E-cadherin negative. Furthermore, seven of eight infiltrated lymph nodes of squamous cell carcinomas were E-cadherin negative. These data indicate that the loss of the cell adhesion molecule E-cadherin is an important factor in the progression of head and neck squamous cell carcinomas, i.e. that down-regulation of expression is associated with de-differentiation and invasion of the tumour cells *in vivo*.

The progressive loss of the capacity of carcinomas to differentiate has been extensively studied in a mouse skin carcinogenesis model. Cano and colleagues have analysed the expression of E- and P-cadherin in mouse skin tumours (Navarro et al 1991). Complete absence of E-cadherin mRNA and protein was found in the spindle cell carcinomas and in three derived cell lines, which all grew tumours rapidly in nude mice. However, E-cadherin was expressed at high levels in the differentiated squamous cell carcinomas and in several epithelioid cell lines, which all formed slowly growing tumours in nude mice. The expression of P-cadherin did not correlate with the degree of differentiation. Interestingly, transfection of HaCa4 cells with an E-cadherin expression vector led to the suppression of tumorigenicity (Navarro et al 1991). E-cadherin expression has been examined in 31 human basal cell carcinomas (Pizarro et al 1993). A characteristic feature of basal cell carcinomas is their low metastatic potential; however, local invasiveness is observed in some cases and these patients exhibit a higher rate of recurrency. E-cadherin was found to be preserved in all superficial basal cell carcinomas in which tumour nests are attached to the undersurface of the epidermis with little penetration into the dermis. Nodular basal cell carcinomas also showed preserved E-cadherin expression; in these, closed tumour masses of various sizes are embedded in the dermis. However, in most cases of infiltrating basal cell carcinoma, E-cadherin expression was clearly reduced, suggesting that this property may be related to infiltrating growth (Pizarro et al 1993).

Moll et al (1993) have recently examined E-cadherin expression on frozen and paraffin sections of 85 human primary infiltrating breast carcinomas. A different picture emerged from the analysis of ductal and lobular breast carcinomas. Forty-one well- and moderately-differentiated infiltrating ductal carcinomas generally showed strong linear staining at the cell borders, at similar levels to luminal cells of normal mammary glands. Twenty-six poorly differentiated, more highly malignant ductal carcinomas were also positive for E-cadherin, although a higher proportion showed reduced staining which was heterogeneous and dotted over the cell borders. In contrast, 16 infiltrating lobular carcinomas (which were either of the dispersed or solid type) did not express E-cadherin, while two cases showed weak staining. *In situ* lesions of lobular carcinomas (three cases) were also E-cadherin negative. The data indicate that loss of E-cadherin expression is an early event in the formation of the lobular but not the ductal type of breast carcinoma. This finding suggests a different mode of invasion in these two breast cancer types. At the invasion front of ductal breast carcinomas, E-cadherin expression is often retained and invasive tumour cells still form truly cohesive (though often thin and thread-like) epithelial sheets. In contrast, lobular breast carcinomas typically invade the connective tissue in a more diffuse manner which may be reflected by the higher rate of recurrency in the patients (Moll et al 1993). Similar results were obtained by Gamallo et al (1993): grade 1 breast carcinomas showed higher E-cadherin immunoreactivity than grade 2 and 3 tumours. Lobular breast carcinomas (seven cases) were all E-cadherin negative. Oka et al (1993) examined 120 breast carcinomas for E-cadherin expression: 47% preserved E-cadherin and were mostly non-differentiated and non-invasive, whereas 53% showed reduced or negative expression and were more highly invasive and poorly differentiated. Reduced E-cadherin expression also correlated with extensive lymph node metastasis (in 74% of the cases) and distant metastasis (in 86% of the cases).

E-cadherin expression in colon cancer has been examined by several laboratories with respect to grade of differentiation, Dukes stage and occurrence of metastasis. Dorudi et al (1993) investigated paraffin sections of 72 colorectal carcinomas: out of 44 well- and moderately differentiated tumours, 36 were positive for E-cadherin, whereas 24 of 28 poorly differentiated carcinomas were E-cadherin negative. Thirty-two of 36 Dukes A and B tumours were E-cadherin positive, while 29 of 36 Dukes C1 and C2 tumours were negative. Furthermore, 20 of 32 lymph node metastases were E-cadherin-negative as were seven of eight liver metastases. Kinsella et al (1993) examined 40 patients with colorectal cancer: 12 had lymph node involvement and nine of these showed reduced E-cadherin expression.

Bussemakers et al (1992) have studied intensively the possible role of E-cadherin expression in both human and rat prostate carcinomas. Several transplantable rat cell lines and the corresponding tumours (from the Dunning R-3327 system) were examined: all non-invasive cell lines which formed tumours with some degree of differentiation expressed E-cadherin mRNA, whereas poorly differentiated lines and their anaplastic tumours were E-cadherin deficient. In a newly arisen, highly metastatic cell line, the decrease in E-cadherin expression occurred concomitantly with the acquisition of metastatic capability, indicating that down-regulation of E-cadherin expression is a prerequisite for invasion to occur. Ninety-two human prostate carcinomas and eight metastases were also analysed, and a remarkable correlation between abnormal E-cadherin staining and increasing grade of the tumour was found (Umbas et al 1992). Within the group of tumours with an intermediate Gleason score, two subclasses may be discriminated, normal or not. It will be interesting to follow up the predictive value of these findings in terms of patients' survival.

Böhm et al (1993) have studied E-cadherin expression in 52 lung carcinomas: E-cadherin expression was reduced or patchy in 14 of 24 moderately differentiated, and reduced or absent in seven of nine poorly differentiated squamous cell carcinomas. All lymph node metastases had down-regulated E-cadherin expression as well as all small-cell lung carcinomas tested. In contrast, only three of 13 moderately and poorly differentiated adenocarcinomas showed reduced E-cadherin expression. In lung carcinomas of the adenosquamous type, the adenoid parts exhibited normal staining whereas the squamous parts showed reduced, patchy expression. Brabant et al (1994) examined the expression of E-cadherin in various types of thyroid carcinomas by Northern blotting and immunofluorescence: in anaplastic thyroid carcinomas ($n=6$) E-cadherin expression was very low or lacking. In papillary carcinomas ($n=23$) E-cadherin mRNA levels varied from nearly normal to greatly reduced, correlating roughly with the protein levels. In follicular carcinomas ($n=9$), E-cadherin mRNA levels were generally high but the immunostaining varied considerably. A few papillary carcinomas lacked immunoreactive E-cadherin despite high mRNA levels. Hürthle cell tumours (four adenomas and two carcinomas) expressed reduced E-cadherin and the protein was localized intracellularly rather than at the cell surface. Otto et al (1994) examined the expression of E-cadherin and of the receptor of autocrine motility factor (gp78) on sections of 123 bladder cancer specimens. Positive expression of E-cadherin and negative expression of gp78 were associated with low risk; only one of 16 patients underwent progression and none died of cancer within 20 months. However, reduction in E-cadherin expression accompanied with increased gp78 expression was associated with poor prognosis; two-thirds of the patients ($n=30$) showed rapid progression and one-third died of cancer. Tohma et al (1992) examined E-cadherin expression in 11 syncytial, 12 transitional and eight fibroblastic meningiomas. E-cadherin was detected in all arachnoid villi (where it was clearly localized in the intermediate junctions) as well as in syncytic and transitional meningiomas. The fibroblastic meningiomas were E-cadherin deficient. Thus, E-cadherin expression seems to be related to the differentiation of meningiomas.

The tumour suppressor gene *APC*: interaction of the *APC* product with β-catenin

Colorectal cancer represents 15% of all cancers in the USA. Colorectal adenocarcinomas (90–95% of colon cancers) start out as polypoid lesions which grow into the lumen or into the bowel wall. The multipstep genetic events of colon cancer progression have been well studied: various genetic changes occur, such as mutations in the genes *APC*, *ras*, *p53* and *DCC*, while tumours progress to malignancy. Mutations of the tumour suppressor gene *APC* occur early in the progression of the disease. The inherited disorder, *familial adenomatous polyposis*, is characterized by an early onset of numerous polyps in the colon, which progress to malignancy. *APC* is also altered with high frequency in sporadic colon cancer (see Fearon & Vogelstein 1990 for a review). Little is known about the function of *APC* in the development of colonic epithelia, nor is it known why the absence of the *APC* gene product promotes colon cancer. The finding that the large APC protein associates with the adherens junction proteins α- and β-catenin (Rubinfeld et al 1993, Su et al 1993) therefore received great attention. How the interaction between membrane-bound E-cadherin, the catenins and APC is organized on a molecular level remained unclear.

To examine the interaction of APC, the catenins and E-cadherin, we isolated the human β-catenin cDNA (Hülsken et al 1994) from a placental cDNA library using the *Xenopus* β-catenin cDNA as a probe (McCrea et al 1991). The deduced amino acid sequence is highly similar to the *Xenopus* sequence (97.4% identity) and is identical to the mouse sequence except for one amino acid difference at position 706 (McCrea et al 1991, Butz et al 1992). Thirteen internal repeats of about 40 amino acids, as originally described for *armadillo* (Peifer & Wieschaus 1990), can be identified, which are flanked by unique N- and C-terminal domains of 131 and 86 amino acids, respectively. The repeat region is highly conserved between human β-catenin and *armadillo* (78% identity), while the N- and C-terminal domains are less conserved (55% and 28%, respectively). A detailed analysis of the repeat sequences revealed that the individual repeat units exhibit various degrees of identity (Fig. 2, Hülsken et al 1994).

Deletion mutants of the human β-catenin cDNA were constructed (Hülsken et al 1994) that encode proteins which lack either or both of the C- and N-terminal domains, or which have truncations within the region of the *armadillo*-like repeats (Fig. 3). The cDNA constructs were transiently expressed in Neuro-2a (mouse neuroblastoma) cells, which express neither E-cadherin nor catenins, and the associations between the various proteins were examined by co-immunoprecipitation. The data allow us to model the multiple interactions of β-catenin

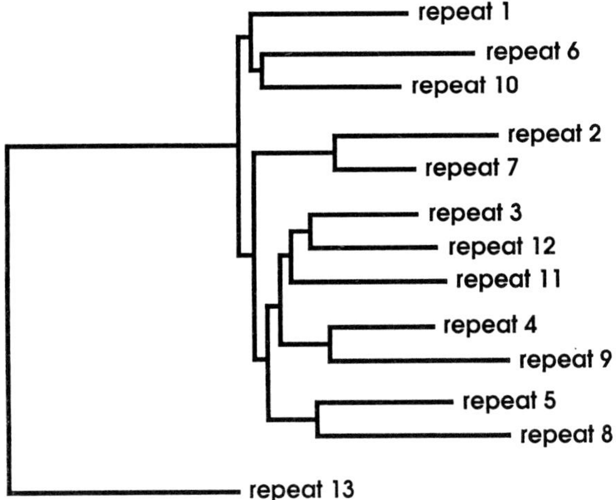

FIG. 2. Dendrogram of the sequence similarity of individual *armadillo* repeats of human β-catenin. The length of horizontal connections indicates the relative distance of sequences. This multiple alignment was created using the TREE algorithm of the HUSAR package (EMBL, Heidelberg, Germany).

within adherens junctions and the APC complex: β-catenin directly interacts with either the cell adhesion molecule E-cadherin or the tumour suppressor gene product APC, and APC and E-cadherin actually compete for binding to the region of the internal, *armadillo*-like repeats of β-catenin (Fig. 4). Plakoglobin (γ-catenin), which is a second member of the *armadillo* family (McCrea et al 1991, Butz et al 1992), forms identical interactions. Furthermore, β-catenin and plakoglobin are direct linkers of both E-cadherin and APC to the cytoskeleton-associated protein α-catenin. This link requires a distinct region of β-catenin, the N-terminal domain. β-catenin and plakoglobin are thus centrally located in both the cell adhesion and the APC complexes; it is also evident that the overall structural arrangement of the E-cadherin/catenin and the APC/catenin complexes is strikingly similar (Fig. 1, Hülsken et al 1994).

APC and E-cadherin are both involved in the suppression of neoplastic transformation and invasiveness (see above); we might therefore speculate that both components cooperate in this function via the competitive interaction with β-catenin. It has been suggested that β-catenin modulates E-cadherin-mediated cell adhesion during processes that require reduced cell adhesion, such as cytokinesis and cell movement, and tyrosine phosphorylation of β-catenin interferes with E-cadherin function (Behrens et al 1993, Hamaguchi et al 1993, Shibamoto et al 1994). APC could thus sequester negative regulatory variants of β-catenin and prevent their interaction with E-cadherin. Alternatively, APC might directly interfere with cell adhesion, e.g. by promoting shedding of epithelial cells into the lumen of the intestine and preventing retention of proliferating cells

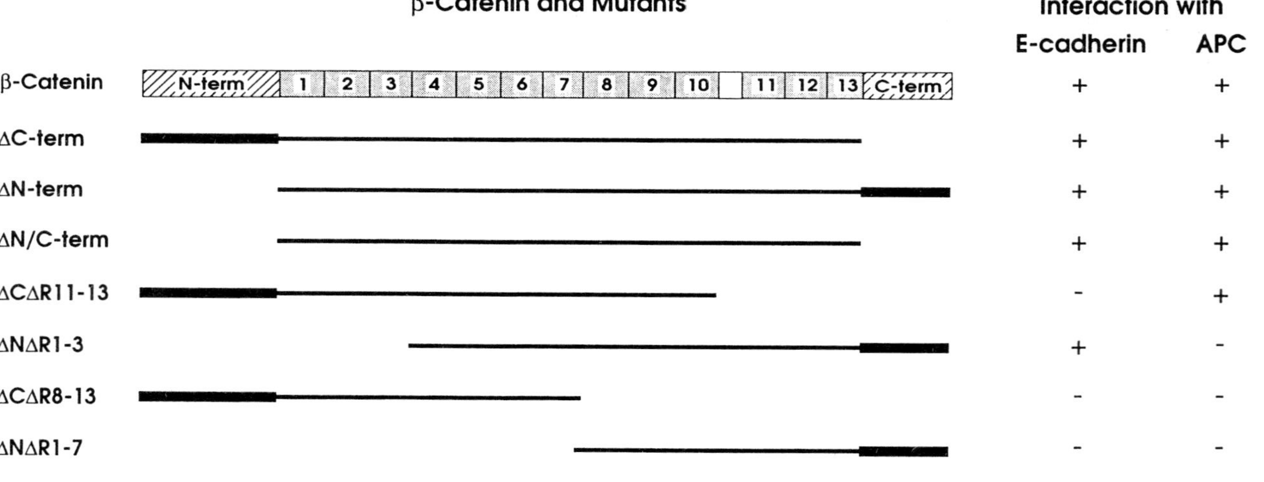

FIG. 3. Schematic representation of full-size and mutant β-catenin and summary of their ability to interact with E-cadherin or APC. On the left, the N- and C-terminal domains and the 13 internal *armadillo*-like repeats of β-catenin are indicated by striped and shadowed boxes, respectively. The following amino acid residues were deleted in the mutated proteins: ΔC-term, 696–781; ΔN-term, 1–131; ΔN/C-term, 1–131 and 696–781; ΔCΔR11–13, 555–781; ΔNΔR1–3, 1–258; ΔCΔR8–13, 424–781; ΔNΔR1–7, 1–422. On the right, a summary of the binding studies between the β-catenin mutants and E-cadherin or APC is presented.

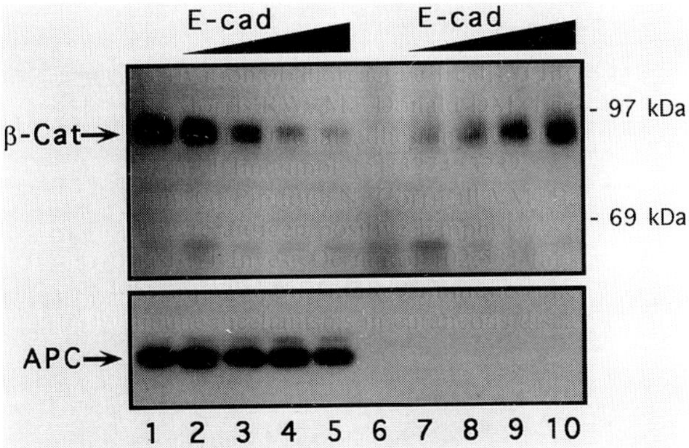

FIG. 4. Competition between E-cadherin and APC for binding to β-catenin. Neuro-2a cells were transfected with 15 μg expression vectors for each of APC and β-catenin plus increasing amounts of expression vector for E-cadherin (lanes 1 and 6, 0 μg; lanes 2 and 7, 1 μg; lanes 3 and 8, 2 μg; lanes 4 and 9, 5 μg; lanes 5 and 10, 10 μg). APC (lanes 1 to 5) and E-cadherin (lanes 6 to 10) were subjected to immunoprecipitation followed by Western blotting for β-catenin (upper part) and APC (lower part).

in the crypts (cf. Smith et al 1993). It is generally accepted that E-cadherin regulates cytoskeletal organization through its interaction with the catenins (Nagafuchi et al 1988, Ozawa et al 1989). The structural similarity of the E-cadherin and APC complexes (Fig. 1) also indicates that APC exerts its tumour suppressor function through modulation of the cytoskeleton.

Plakoglobin (γ-catenin) interacts equally with either the cell adhesion molecule E-cadherin or the tumour suppressor gene product APC. β-catenin and plakoglobin might thus be functionally redundant; this question can now be examined in systems where the gene for one or the other is deleted. Alternatively, β-catenin and plakoglobin might have functionally different roles and may be subject to separate regulation. It is known that both proteins differ in their association with cell junctions *in vivo*, i.e. β-catenin is largely associated with adherens junctions whereas plakoglobin is also part of desmosomes (Cowin et al 1986).

Recent evidence suggests that β-catenin, besides functioning in adhesion complex formation and in the interaction with APC, is also involved in signal transduction: its homologue *armadillo* is part of a signal cascade from *wingless* to *engrailed* in determining segment polarity in *Drosophila* (Bejsovec & Wieschaus 1993, van Leeuwen et al 1994, Bradley et al 1993, Hinck et al 1994) and microinjection of anti-β-catenin antibody into early *Xenopus* embryos results in a duplication of the embryonic axes (McCrea et al 1993). It remains to be determined whether these signalling processes make use of β-catenin in a membrane- or APC-associated complex or whether independent interactions occur.

Acknowledgements

Our work is supported by the Deutsche Krebshilfe (Mildred Scheel-Stiftung) and the Bundesministerium für Forschung und Technology (BMFT).

References

Becker KF, Atkinson MJ, Reich U et al 1994 E-cadherin gene mutations provide clues to diffuse-type gastric carcinomas. Cancer Res 54:3845–3852

Behrens J, Mareel MM, Van Roy FM, Birchmeier W 1989 Dissecting tumor cell invasion: epithelial cells acquire invasive properties after the loss of uvomorulin-mediated cell–cell adhesion. J Cell Biol 108:2435–2447

Behrens J, Vakaet L, Friis R et al 1993 Loss of epithelial differentiation and gain of invasiveness correlates with tyrosine phosphorylation of the E-cadherin/β-catenin complex in cells transformed with a temperature-sensitive *v-SRC* gene. J Cell Biol 120:757–766

Bejsovec A, Wieschaus E 1993 Segment polarity gene interactions modulate epidermal patterning in *Drosophila* embryos. Development 119:501–517

Birchmeier W, Behrens J 1994 Cadherin expression in carcinomas: role in the formation of cell junctions and the prevention of invasiveness. Biochim Biophys Acta 1198: 11–26

Böhm M, Totzeck B, Birchmeier W, Wieland I 1993 Difference of E-cadherin expression levels and patterns in primary and metastatic human lung cancer. Clin Exp Metastasis 12:55–62

Brabant G, Hoang-Vu C, Cetin Y et al 1994 E-cadherin: a differentiation marker in thyroid malignancies. Cancer Res 53:4987–4993

Bradley RS, Cowin P, Brown AM 1993 Expression of Wnt-1 in PC12 cells results in modulation of plakoglobin and E-cadherin and increased cellular adhesion. J Cell Biol 123:1857–1865

Bussemakers MJC, Van Morselaar RJA, Giroldi LA et al 1992 Decreased expression of E-cadherin in the progression of rat prostatic cancer. Cancer Res 52:2916–2922

Butz S, Stappert J, Weissig H, Kemler R 1992 Plakoglobin and beta-catenin: distinct but closely related. Science 257:1142–1144

Buxton RS, Magee AI 1992 Structure and interactions of desmosomal and other cadherins. Semin Cell Biol 3:157–167

Coman DR 1944 Decreased mutual adhesiveness, a property of cells from squamous cell carcinomas. Cancer Res 1:625–629.

Cowin P, Kapprell HP, Franke WW, Tamkun J, Hynes RO 1986 Plakoglobin: a protein common to different kinds of intercellular adhering junctions. Cell 46:1063–1073

Edelman GM 1988 Morphoregulatory molecules. Biochemistry 27:3533–3543

Fearon ER, Vogelstein B 1990 A genetic model for colorectal tumorigenesis. Cell 61:759–767

Frixen UH, Behrens J, Sachs M et al 1991 E-cadherin-mediated cell–cell adhesion prevents invasiveness of human carcinoma cells. J Cell Biol 113:173–185

Fujimori T, Takeichi M 1993 Disruption of epithelioid cell–cell adhesion by exogenous expression of a mutated nonfunctional N-cadherin. Mol Biol Cell 4:37–47

Gabbert H, Wagner R, Moll R, Gerharz CD 1985 Tumor dedifferentiation: an important step in tumor invasion. Exp Metastasis 3:257–279

Hamaguchi M, Matsuyoshi N, Ohnishi Y, Gotoh B, Takeichi M, Nagai Y 1993 $p60^{v\text{-}src}$ causes tyrosine phosphorylation and inactivation of the N-cadherin–catenin cell adhesion system. EMBO (Eur Mol Biol Organ) 12:307–314

Herrenknecht K, Ozawa M, Eckerskorn C, Lottspeich F, Lenter M, Kemler R 1991 The uvomorulin-anchorage protein alpha catenin is a vinculin homologue. Proc Natl Acad Sci USA 88:9156–9160

Hinck L, Nelson WJ, Papkoff J 1994 Wnt-1 modulates cell–cell adhesion in mammalian cells by stabilizing β-catenin binding to the cell adhesion protein cadherin. J Cell Biol 124:729–741

Hirano S, Kimoto N, Shimoyama Y, Hirohashi S, Takeichi M 1992 Identification of a neural α-catenin as a key regulator of cadherin function and multicellular organization. Cell 70:293–301

Hülsken J, Birchmeier W, Behrens J 1994 E-cadherin and APC compete for the interaction with β-catenin and the cytoskeleton. J Cell Biol, in press

Kemler R, Ozawa M, Ringwald M 1989 Calcium-dependent cell adhesion molecules. Curr Opin Cell Biol 1:892–897

Kinsella AR, Green B, Lepts GC, Hill CL, Bowie G, Taylor BA 1993 The role of the cell adhesion molecule E-cadherin in large bowel tumor cell invasion and metastasis. Br J Cancer 67:904–909

Kintner C 1992 Regulation of embryonic cell adhesion by the cadherin cytoplasmic domain. Cell 69:225–236

Knudsen KA, Wheelock MJ 1992 Plakoglobin, or an 83-kD homologue distinct from β-catenin, interacts with E-cadherin and N-cadherin. J Cell Biol 118:671–679

McCrea PD, Turck CW, Gumbiner B 1991 A homolog of the armadillo protein in *Drosophila* (plakoglobin) associated with E-cadherin. Science 254:1359–1361

McCrea PD, Brieher WM, Gumbiner BM 1993 Induction of a secondary body axis in *Xenopus* by antibodies to β-catenin. J Cell Biol 123:477–484

Moll R, Mitze M, Frixen UH, Birchmeier W 1993 Differential loss of E-cadherin expression in infiltrating ductal and lobular breast carcinomas. Am J Pathol 143:1731–1742

Nagafuchi A, Takeichi M 1988 Cell binding function of E-cadherin is regulated by the cytoplasmic domain. EMBO J 7:3679–3684

Nagafuchi A, Takeichi M, Tsukita S 1991 The 102 kd cadherin-associated protein: similarity to vinculin and posttranscriptional regulation of expression. Cell 65:849–857

Navarro P, Gomez M, Pizarro A, Gamallo C, Quintanilla M, Cano A 1991 A role for the E-cadherin cell–cell adhesion molecule during tumor progression of mouse epidermal carcinogenesis. J Cell Biol 115:517–533

Oda T, Kanai Y, Shimoyama Y, Nagafuchi A, Tsukita S, Hirohashi S 1993 Cloning of the human α-catenin cDNA and its aberrant mRNA in a human cancer cell line. Cancer Res 53:1696–1701

Oda T, Kanai Y, Oyama T et al 1994 E-cadherin gene mutations in human gastric carcinoma cell lines. Proc Natl Acad Sci USA 91:1858–1862

Oka H, Shiozaki H, Kobayashi K et al 1993 Expression of E-cadherin cell adhesion molecules in human breast cancer tissues and its relationship to metastasis. Cancer Res 53:1696–1701

Ozawa M, Baribault H, Kemler R 1989 The cytoplasmic domain of the cell adhesion molecule uvomorulin associates with three independent proteins structurally related in different species. EMBO (Eur Mol Biol Organ) J 8:1711–1717

Ozawa M, Engel J, Kemler R 1990 Single amino acid substitutions in one Ca^{2+} binding site of uvomorulin abolish the adhesive function. Cell 63:1033–1038

Peifer M, Wieschaus E 1990 The segment polarity gene *armadillo* encodes a functionally modular protein that is the Drosophila homolog of human plakoglobin. Cell 63:1167–1176

Piepenhagen PA, Nelson WJ 1993 Defining E-cadherin-associated protein complexes in epithelial cells: plakoglobin, β- and γ-catenin are distinct components. J Cell Sci 104:751–762

Pizarro A, Benito N, Navarro P et al 1993 E-cadherin expression in basal carcinomas. Br J Cancer 69:157–162

Reichmann E, Schwarz H, Deiner EM et al 1992 Activation of an inducible c-FosER fusion protein causes loss of epithelial polarity and triggers epithelial–fibroblastoid cell conversion. Cell 71:1103–1116

Risinger JI, Berchuck A, Kohler MF, Boyd J 1994 Mutations of the E-cadherin gene in human gynecologic cancers. Nature Genet 7:98–102

Rouleau GA, Merel P, Lutchman M et al 1993 Alteration in a new gene encoding a putative membrane-organizing protein causes neurofibromatosis type 2. Nature 363:515–521

Rubinfeld B, Souza B, Albert I et al 1993 Association of the *APC* gene product with beta-catenin. Science 262:1731–1734

Schipper JH, Frixen UH, Behrens J, Unger A, Jahnke K, Birchmeier W 1991 E-cadherin expression in squamous cell carcinomas of head and neck: inverse correlation with tumor dedifferentiation and lymph node metastasis. Cancer Res 51:6328–6337

Shibamoto S, Hayakawa M, Takeuchi K et al 1994 Tyrosine phosphorylation of β-catenin and plakoglobin enhanced by hepatocyte growth factor and epidermal growth factor in human carcinoma cells. Cell Adhesion Commun 4:295–305

Shimoyama Y, Nagafuchi A, Fujita S et al 1992 Cadherin dysfunction in a human cancer cell line: possible involvement of loss of α-catenin expression in reduced cell–cell adhesiveness. Cancer Res 52:1–5

Smith KJ, Johnson KA, Bryan TM et al 1993 The *APC* gene product in normal and tumor cells. Proc Natl Acad Sci USA 90:2846–2850

Su LK, Vogelstein B, Kinzler KW 1993 Association of the APC tumor suppressor protein with catenins. Science 262:1734–1737

Takeichi M 1991 Cadherin cell adhesion receptors as morphology regulators. 251: 1451–1455

Trofatter JA, MacColin MM, Rutter JL et al 1993 A novel moesin-, ezrin-, radixin-like gene is a candidate for the neurofibromatosis 2 tumor suppressor. Cell 72:791–800

Tsukita S, Itoh M, Nagafuchi A, Yonemura S 1993 Submembranous junctional plaque proteins include potential tumor suppressor molecules. J Cell Biol 123:1049–1053

Umbas R, Schalken JA, Aalders TW et al 1992 Expression of the cellular adhesion molecule E-cadherin is reduced or absent in high-grade prostate cancer. Cancer Res 52:5104–5109

van Leeuwen F, Samos CH, Nusse R 1994 Biological activity of soluble wingless protein in cultured *Drosophila* imaginal disc cells. Nature 368:342–344

Vleminckx K, Vakaet L, Mareel M Jr, Fiers W, van Roy F 1991 Genetic manipulation of E-cadherin expression by epithelial tumor cells reveals an invasion suppressor role. Cell 66:107–119

Woods DF, Bryant PJ 1991 The discs-large tumor suppressor gene of Drosophila encodes a guanylate kinase homolog localized at septate junctions. Cell 66:451–464

DISCUSSION

Hynes: It seems to me that the way you have the APC involvement drawn is inconsistent with the simple model of the tumour suppressor gene's product strengthening the adhesion: if APC is competing with E-cadherin for the cytoskeletal components, you would have thought it might undo the junctions and then, when you lose it, adhesion would be better.

Birchmeier: This is the crucial question. One has to think of β-catenin as a kind of modulator of adhesion. On the one hand, it is an essential bridge between E-cadherin and the cytoskeleton; on the other hand, it can be phosphorylated on tyrosines, resulting in reduced adhesion. The idea is that APC or E-cadherin could remove negative (e.g. tyrosine-phosphorylated) variants of β-catenin and by such a mechanism work in the same direction.

Hynes: So, does APC bind better to the phosphorylated form than to the unphosphorylated form of catenin?

Birchmeier: We don't know yet, but these are the experiments we intend to do next. The rumours are that the N-terminus of β-catenin is the tyrosine-phosphorylated part.

Herrlich: You indicated that both E-cadherin and APC bind to β-catenin. Is β-catenin at all limiting? If you transfect an E-cadherin-expressing cell with an excess of APC, do you eliminate cell adhesion?

Birchmeier: When we saw Vogelstein's and Polakis' papers (Su et al 1993, Rubinfeld et al 1993), we quickly cloned the fragment of APC that binds β-catenin. We are currently looking at whether the APC fragment interferes with cell adhesion in a dominant negative fashion. In the intestine, APC might in fact interfere with cell adhesion, e.g. promote shedding of epithelial cells into the lumen and prevent retention of proliferative cells in the crypts.

Riethmüller: It is often assumed that epithelial cells are immobile. I wonder whether the view of the epithelium as a cobblestone texture, in which the position of every stone is permanently fixed, is correct. For instance, in the small intestine, the epithelial cells move rapidly from the crypt to the tip of the villus. During this process, assembly and disassembly of desmosomes must occur. I view epithelial cells as very mobile, at least in the gut.

Birchmeier: It's all relative: in comparison to fibroblasts, epithelial cells are still fairly immobile.

Riethmüller: But, in well differentiated metastases, E-cadherin expression often returns, as we have seen. E-cadherin down-regulation may be a transient phenomenon, but if metastatic nodules form again at an epithelial parenchyma, you have perfect tight junctions.

Birchmeier: E-cadherin expression is in fact reinstated in many carcinomas; you might explain invasiveness as being a consequence of the transient down-regulation of E-cadherin. However, other molecules, such as scatter factor, do not induce down-regulation of E-cadherin yet still promote invasiveness. I'm actually waiting for someone to discover cell types which normally lack E-cadherin but which express it in metastasis. Remember that Langerhans' cells suddenly express E-cadherin in the skin. No one seems to have looked at melanomas, but it's known that the most aggressive melanomas can express keratins and I would not be surprised if they also begin to express E-cadherin.

Stanley: Amin Tang and Barbara Gilchrest at Boston University have reported that normal melanocytes express E-cadherin (Tang et al 1994). This may have something to do with how they maintain their attachment to the epidermis. They've looked at various melanoma cell lines: they tend to lose their E-cadherin as well.

Hynes: This idea of down-regulating and then reinstating expression of the same adhesion molecule is exactly what is thought to happen with neural crest cells in development: cadherin expression falls as they exit the neural tube, it is absent during the time of migration and it turns on again as they aggregate into ganglia. So I think the idea that cells can temporarily lose adhesion is a very reasonable one.

Stanley: In the ductal carcinoma tumour and other tumours that you have looked at that express E-cadherin, have you looked for the absence of α-catenin, which has also been reported in E-cadherin-expressing tumours?

Birchmeier: We haven't yet, but other groups have. Hirohashi's group recently reported the absence of α-catenin in E-cadherin-positive ductal breast carcinomas (Ochiai et al 1994).

Stanley: You've looked at the parts of β-catenin that are necessary for binding: have you looked at the parts of the cytoplasmic domain of E-cadherin that are necessary for binding plakoglobin and β-catenin, and are they the same?

Birchmeier: Presumably you are asking that question because you think that plakoglobin might bind to a different region on the desmosomal cadherins. We have not studied this in detail; we just assume that the sites are the same because β-catenin and plakoglobin are so similar and compete for binding. Neither have we looked at the region of plakoglobin which binds to α-catenin—again, we just assume that it's the N-terminus. We don't really know whether plakoglobin has a function in adherens junctions: perhaps the plakoglobin and β-catenin knockout mice will tell us more.

Stanley: Why does binding to the cytoskeleton increase the adhesive function of the extracellular domain of E-cadherin?

Birchmeier: Cadherins are very strong adhesion molecules. In the case of N-CAM, for instance, you can reconstitute adhesion in lipid vesicles and then perturb this adhesion with antibodies—you could never do this with cadherins. Apparently, with cadherins, interaction with the cytoskeleton is necessary for strong adhesion to occur.

Stanley: Do you think cadherins mediate a signal across the transmembrane domain that changes the conformation of the extracellular domain?

Hynes: There are two things going on; part may be due to a conformational change, but there is a general feeling in the integrin field (and I think this is also true for cadherins) that part of the role of the cytoskeleton is to collect adhesion receptors into patches, so that instead of having a single-point contact, you've got a multipoint contact, which is stronger. Even if they're coming on and off all the time, some of them are always on. You can show by careful measurements that if you cluster the integrins into patches, the adhesion is stronger.

Birchmeier: I agree. It is also unlikely that EGF signals to the EGF receptor kinase in the form of the monomer. It works through oligomerization of receptors. The same is probably true for inside–out and outside–in signalling of E-cadherins.

In the case of the tyrosine kinases, there is also autophosphorylation of the kinase domain that could cause an additional change in conformation.

Hynes: There could be other things going on, but you can show that by clustering adhesion molecules you increase the strength of the cell adhesion.

Stanley: These molecules could still have a means for clustering other than by means of the cytoskeleton. For instance, they might cluster through direct homophilic interactions of their cytoplasmic tails.

Garrod: Is there any evidence that you need an intact cytoskeleton for E-cadherin to function? The reason I ask is because in keratin knockouts you can get perfectly good desmosome formation without any keratin cytoskeleton. I think the evidence that you have to have β-catenin binding and α-catenin binding is very strong, but I'm not sure whether that actually means that you need an intact cytoskeleton.

Birchmeier: There are still missing links between α-catenin and the actin cytoskeleton.

Sonnenberg: Can the adherent junctions still assemble in the presence of cytochalasin D or B?

Birchmeier: I guess not.

Wagner: Mark Ginsberg has results from experiments with GPIIb/IIIa that show that the deletion of the α_{IIb} cytoplasmic domain activates the molecule (O'Toole et al 1991).

Ruggeri: The effect you refer to may not be the consequence of altering linkage to the cytoskeleton. I guess the question is specifically to do with the cytoskeleton.

Hynes: Both of these things (i.e. activation of the integrins themselves, as well as clustering) could be going on—they probably are.

Wagner: The question was: could there be a transduction of a conformational change across the membrane?

Ruggeri: But not necessarily through the cytoskeleton connection—it could be another intra-cytoplasmic protein that interacts with the cytoplasmic tail of the receptor.

Hynes: For the integrins, the evidence is that the β-subunit binds to the cytoskeleton: deletion of this leads to defective adhesion (Hynes 1992). There are other situations where deletions in the cytoplasmic domain actually activate the receptor. The tidiest model, although it doesn't explain everything, is that the α cytoplasmic domain is acting as a negative regulator of the β cytoplasmic domain. Again, there's probably more going on than this.

Ruggeri: Some people invoke the presence of a putative integrin-activating complex. This is an as yet unidentified regulatory molecule that could interact with the cytoplasmic tails.

Sonnenberg: Integrins do not function in the presence of cytochalasin B, thus evidently they need to make connections with the actin cytoskeleton for clustering, which is necessary for their function. I assume that this will also be true for the cadherins.

Birchmeier: In the integrins, is the α-subunit the negative regulator of the β?

Hynes: The β-subunit is constitutively functional. If you put a β cytoplasmic domain on some random reporter transmembrane protein, the chimera will go to a cytoskeletal location.

Wagner: I don't think anything is black and white: some integrins, like GPIIb/IIIa, even function when coated onto beads.

Koopman: Is the expression of E-cadherin also affected by its adhesive potential? When you shut its adhesive function off, for instance with scatter factor, does its pattern of expression change?

Birchmeier: The level of E-cadherin expression does not change, but its distribution does. The cell also changes shape, i.e. somehow adhesion is impaired.

Koopman: So it's not shut off temporarily and then re-expressed as soon as cells meet again.

Etzioni: What is the significance of measuring the expression of these adhesion molecules and looking at mutations in the development of cancer considering the fact that pathologists know already what is going on when looking at the differentiation state?

Birchmeier: I get asked this question a lot: pathologists would like to be able to look at the differentiation state of the tumour and then know what the prospects of the patients are. We would like to know exactly why, in each type of tumour, the tumours progress. I hope that in the future one will treat patients who, for instance, have a defect in E-cadherin by mutation differently from patients who have down-regulation in E-cadherin expression. We will have to know exactly what is wrong at the molecular level in different carcinomas to come up with an effective therapy.

Barker: We've looked at skin squamous cell carcinomas and shown that the undifferentiated tumours express less E-cadherin than the well differentiated tumours. But I keep wondering whether we're actually looking at the right thing: wouldn't it be more relevant to look at the metastases and see what the pattern of cadherin expression is from the metastatic deposits? Have you done that with prostate and breast cancers, for instance?

Birchmeier: We have just done one study on head and neck carcinomas. We looked at the lymph node metastases, and in seven out of eight cases E-cadherin was no longer expressed, irrespective of the state of E-cadherin expression in the primary tumour (Schipper et al 1991). It looked to us as if E-cadherin-negative cells of the primary tumour had migrated to the lymph nodes. We are currently doing knockout experiments looking at E-cadherin heterozygous (+/−) mice, and then inducing tumours by different means. It will be interesting to see whether we observe loss of heterozygosity preferentially in the metastases. I should again mention the interesting report by Becker et al (1994) on gastric carcinomas, where they find skipping of exons eight or nine, i.e. the E-cadherin is still there, but one of the two exons is deleted. This experiment suggests that you need a little bit of E-cadherin left for development of the tumour.

Herrlich: E-cadherin can occur in an active or an inactive conformation and phosphorylation can influence its conformation. Do any of your antibodies distinguish between functional (adhesive versus non-adhesive) states?

Birchmeier: We have antibodies which are functionally active, but they do not differentiate between active and inactive states of E-cadherin in immunofluorescence experiments. Our anti-Arc-1 antibody binds to the adherens junction regions of fixed cells, but this reaction is destroyed if EGTA is added. So, this particular antibody binds to E-cadherin when complexed with Ca^{2+}.

Parekh: You said that 50% of gastric cancers have this mutation in the E-cadherin and weak junctions. Do the 50% that don't have this mutation still have E-cadherin in the adherens junction?

Birchmeier: We have collaborated with Gert Riethmüller's lab (cf. Mayer et al 1993) looking at down-regulation of E-cadherin in gastric carcinomas. E-cadherin expression is often down-regulated; in the Höfler/Becker cases E-cadherin is not down-regulated. So I guess that the ratio of cases with down-regulation:mutation is approximately 50:50.

References

Becker KF, Atkinson MJ, Reich U et al 1994 E-cadherin gene mutations provide clues to diffuse-type gastric carcinomas. Cancer Res 54:3845–3852

Hynes RO 1992 Integrins: versatility, modulation, and signaling in cell adhesion. Cell 69:11–25

Mayer B, Johnson JP, Leitl F et al 1993 E-cadherin expression in primary and metastatic gastric cancer—down regulation correlates with cellular dedifferentiation and glandular disintegration. Cancer Res 53:1690–1695

Ochiai A, Akimoto S, Shimoyama Y, Nagafuchi A, Tsukita S, Hirohashi S 1994 Frequent loss of α-catenin expression in scirrhous carcinomas with scattered cell growth. Jpn J Cancer Res 85:266–273

O'Toole TE, Mandelman D, Forsyth J, Shatil SJ, Plow EF, Ginsberg MH 1991 Modulation of the affinity of integrin $\alpha_{IIb}\beta_3$ by the cytoplasmic domain of α_{IIb}. Science 254:845–847

Rubinfeld B, Souza B, Albert I et al 1993 Association of the *APC* gene product with β-catenin. Science 262:1731–1734

Schipper JH, Frixen UH, Behrens J, Unger A, Jahnke K, Birchmeier W 1991 E-cadherin expression in squamous cell carcinoma of head and neck: inverse correlation with tumor differentiation. Cancer Res 51:6328–6337

Su L-K, Vogelstein B, Kinzler KW 1993 Association of the APC tumour suppressor protein with catenins. Science 262:1734–1737

Tang A, Eller MS, Hara M, Yaar M, Hirohashi S, Gilchrest BA 1994 E-cadherin is the major mediator of human melanocyte adhesion to keratinocytes *in vitro*. J Cell Sci 107:983–992

The role of CD44 splice variants in human metastatic cancer

Jonathan Sleeman, Jürgen Moll, Larry Sherman, Peter Dall, Steven T Pals*, Helmut Ponta and Peter Herrlich

*Kernforschungszentrum Karlsruhe, Institut für Genetik, Postfach 3640, D-76021 Karlsruhe, Germany and *Academic Medical Centre, Department of Pathology, Meibergdreef 9, NL-1105 AZ Amsterdam, The Netherlands*

Abstract. The large family of CD44 splice variants are likely to serve multiple functions in the embryo and in the adult organism. This is reflected in their complex patterns of expression. In molecular terms these functions are largely unknown. Certain splice variants (CD44v) can promote the metastatic behaviour of cancer cells. In human colon and breast cancer the presence of epitopes encoded by exon v6 on primary resected tumour material indicates poor prognosis. Metastasis-promoting splice variants differ from those that seem not to have a role in the induction of metastasis by the formation of homomultimeric complexes in the plasma membrane of cells. This may increase their affinity to ligands such as hyaluronate. The affinity can be further regulated over a range from low to very high by cell-specific modification. The fact that CD44v epitopes are found on normal epithelial cells such as skin, cervical epithelium and bladder enforces cautious evaluation of the significance of CD44v expression in human cancer. Nevertheless, certain epitopes can serve as tools in early diagnosis of certain cancers and will facilitate the development of specific targeted therapy.

1995 Cell adhesion and human disease. Wiley, Chichester (Ciba Foundation Symposium 189) p 142–156

Cells that grow and/or migrate require specific cell surface properties that destroy obstacles and promote contact as they move and subsequently attach to new sites. Cell surface-bound enzymes such as gelatinase, stromelysin or collagenase type I induce the degradation of their specific substrates from the extracellular matrix. On one hand, adhesion molecules mediate long-lasting contact to the extracellular matrix and other cells, while on the other hand they mediate the directional movement peculiar to certain cell types. It is likely that the process of metastasis involves several of these mechanisms which are used during normal cellular migration.

Our understanding of the mechanisms by which cell adhesion molecules work is still in its infancy. Although there must be enormous specificity in cell–cell and cell–matrix recognition, we are just beginning to understand molecular details of the action of specific molecules. This volume gives a number of examples. Interestingly, several adhesion molecules occur in a large variety of different isoforms, e.g. the integrins with their combinatorial association of one of many α subunits with one of many β subunits (reviewed by Hynes 1992). The adhesion molecules designated CD44 represent another highly variable family of isoforms. These are generated by alternative splicing (reviewed in Herrlich et al 1993). The isoforms have putative functions in development, somatic cell functions and metastatic disease.

The CD44 splice variants and their modifications

The designation CD44 classifies a family of proteins that share the N-terminal (extracellular) and C-terminal (transmembrane, cytoplasmic) sequences. They differ in the central, extracellular part by use of ten 'variant' exons (v1 to v10) that are either completely excised (CD44 standard form, CD44s) or used in various combinations (CD44 variants, CD44v) (Screaton et al 1992, Tölg et al 1993). More than 30 different splice products have been detected by PCR analysis (van Weering et al 1994). Although it is not clear whether all splice products of CD44 are translated into surface proteins, sequence diversity exceeds that of any other known protein family created by alternative splicing.

In addition to sequence diversity, the CD44 proteins are post-translationally modified (reviewed in Lesley et al 1993). There is evidence that the modification contributes to functional variability and that modification depends on cell type and culture conditions (reviewed by Lesley et al 1993). The cytoplasmic C-terminal tail carries several serines that can be phosphorylated (Camp et al 1992). The extracellular domains are glycosylated (reviewed in Lesley et al 1993). While the smallest CD44 isoform (35 kDa according to amino acid composition) is usually detected as the 85 kDa form largely as a result of glycosylation, some isoforms can migrate with an apparent molecular mass of 200 kDa and more (reviewed by Lesley et al 1993). The CD44 family of proteins is likely to exhibit a large variety of functionally different specificities despite the fact that individual domains, e.g. the cytoplasmic tail and the N-terminus, are shared.

Expression pattern

The smallest CD44 protein (CD44s) is expressed on many different tissues. Antibodies to the N-terminus of CD44 stain many tissues of the embryo as well as of the adult, including a variety of non-epithelial cells (Heider et al 1993a, Wheatley et al 1993). Many cells of haemopoietic origin carry CD44, as do cells undergoing proliferative expansion, e.g. in normal wound healing, at the epiphyseal growth plate of long bones (C. B. Underhill, personal communication) and at the base of intestinal crypts (Wirth et al 1993). Most tumour cells also stain with such anti-CD44s antibodies. Because the N-terminal domain carries a hyaluronate binding motif (Culty et al 1990, Aruffo et al 1990), at least part of the pattern of expression is probably related to the distribution of hyaluronate. Hyaluronate is the major 'space-holding' compound used by cells to 'reserve' space as they proliferate and migrate. Hyaluronate receptors such as CD44 are believed to be part of the machinery by which cells recognize hyaluronate for these processes.

The distribution of individual CD44 splice variants that carry specific variant exon sequences is far more restricted (Heider et al 1993a), suggesting that these larger isoforms recruit different additional functions. Epitopes encoded by variant exons are often found transiently on embryonic or adult epithelia, on certain cells in the central and peripheral nervous system and on haemopoietic cells at specific stages of development. (A summary of the expression of CD44 is presented by Heider et al 1993a and Wirth et al 1993.)

CD44 variants promoting metastatic growth of tumours

Animal and human cancers with metastatic properties tend to express increased levels and new larger variants of CD44 on their surface. Furthermore, non-metastatic cell lines become metastatic upon transfection with expression clones carrying specific CD44 variant cDNAs (e.g. CD44v4–v7 or CD44v6,7) (Günthert et al 1991, Rudy et al 1993). Anti-CD44v6 antibodies injected intravenously into syngeneic rats immediately after subcutaneous injection of CD44v4–v7-expressing tumour cells could prevent the outgrowth of lymph node and lung metastases, indicating that the promoting effect of CD44v4–v7 was required early in the spread to draining lymphatic tissue (Reber et al 1990, Seiter et al 1993).

Considering the enormous structural diversity of CD44, it would be amazing if screening of individual CD44 epitopes in human cancers yielded an easily interpretable result with regard to the role of CD44 in tumour progression. Nevertheless, several laboratories have attempted to relate the presence of epitopes to the phenotype of the tumour. A survey of these results is given in Table 1. Several features can be provisionally deduced. In neuroblastoma, Burkitt's lymphoma and melanoma, CD44 variants often appear to be absent. In one study, the absence of CD44s correlated with poor survival in patients with neuroblastoma (Favrot et al 1993). This indicates that tumours can use pathways of metastasis that are unrelated to the presence of a CD44 molecule. Tumours originating from epithelia that normally carry CD44 variant epitopes on their surface remain positive for CD44v and often exhibit enhanced levels of CD44 transcripts (Heider et al 1993b, Dall et al 1994). Whether elevated transcription is connected with tumour progression is not yet clear. Such cancers may, however, carry specific epitopes (neo-epitopes) not expressed in normal cells, e.g. cervical cancer is stained with a monoclonal antibody whose epitope is jointly formed by exons v7 and v8. The epitope first appears at the carcinoma *in situ* stage of cervical cancer (Dall et al 1994). Perhaps the most interesting

TABLE 1 Expression of CD44 isoforms in human cancer

Tumour	*Surface expression of CD44 epitope*[a]	*RT-PCR evidence for CD44 variants*	*Positive (+) or negative (−) correlation with tumour progression stage and/or clinical outcome*	*Number of samples*	*Reference*
Breast	v3, v5, v6, v7/8, v10 (+)	ND	survival (−) (v3, v5, v6)	100	Kaufmann et al 1944
	v3, v5, v6, v7/8, v10 (+)	v3–v10, v4–v7, v8–v10, complex pattern	progression (+) (v3–v10)	24	our unpublished results
	v3, v4/5, v6, v8/9 (+)	ND	ND	10	Fox et al 1993
	ND	v3–v10, complex pattern	ND	24	Matsumara & Tarin 1992
Gastric	v5, v6 (+)	v3–v10, v5, v6	ND	42	Heider et al 1993b
	v9 (+)	ND	survival (−)	31	Mayer et al 1993
Colon	v6 (+)	v6–v10	progression (+)	38	Wielenga et al 1993
	v3, v4/5, v6 (+)	ND	ND	5	Fox et al 1993
	ND	v3–v10, complex pattern	ND	10	Matsumara & Tarin 1992
	ND	v3–v10, complex pattern	progression (+)	24	Finn et al 1994
Non-Hodgkin lymphoma	v6 (+)	ND	progression (+)	36	Koopman et al 1993
Skin (squamous cell)	v6 (−)	v6 (−)	ND	30	Salmi et al 1993
Cervix (squamous cell)	v7/8 (+)	v3–v10, v8–v10, v4–v7, complex pattern	ND	17	Dall et al 1994
Glioblastoma	no epitopes	no variants	ND	17	Li et al 1993
Neuroblastoma	CD44s (+)	ND	survival (+)	52	Favrot et al 1993

[a](+)/(−) indicates increased/decreased CD44 epitope expression compared with normal tissue.
ND, not determined.

group of tumours arises from CD44 variant epitope-negative tissues. Aggressive non-Hodgkin lymphoma (Koopman et al 1993), Dukes C/D colorectal carcinoma (Wielenga et al 1993) and invasive breast tumours (Kaufmann et al 1994) carry a CD44 epitope encoded by exon v6. The presence of this epitope on primary tumours predicts poor survival of the patients (Kaufmann et al 1994, and Fig. 1). Multivariate analyses define the epitope as an independent prognosticator (independent of Dukes classification, of erbB2 and epidermal growth factor receptor [EGFR] amplification or other standard parameters). The lifespan of cancer patients is predominantly limited by the formation of distant metastases. The correlation of an epitope with poor survival therefore supports the idea that this epitope plays a role in the generation of metastases.

The expression of several CD44 variant epitopes, e.g. an epitope encoded by exon v5 in colorectal tumours (Wielenga et al 1993) or of v10 in mammary carcinomas (Kaufmann et al 1994), does not correlate with poor survival, and certain isoforms of CD44, e.g. the smallest ubiquitously expressed CD44s form, cannot induce metastatic behaviour if over-expressed in non-metastatic cell lines (Günthert et al 1991). CD44s has, however, been shown to help in the formation

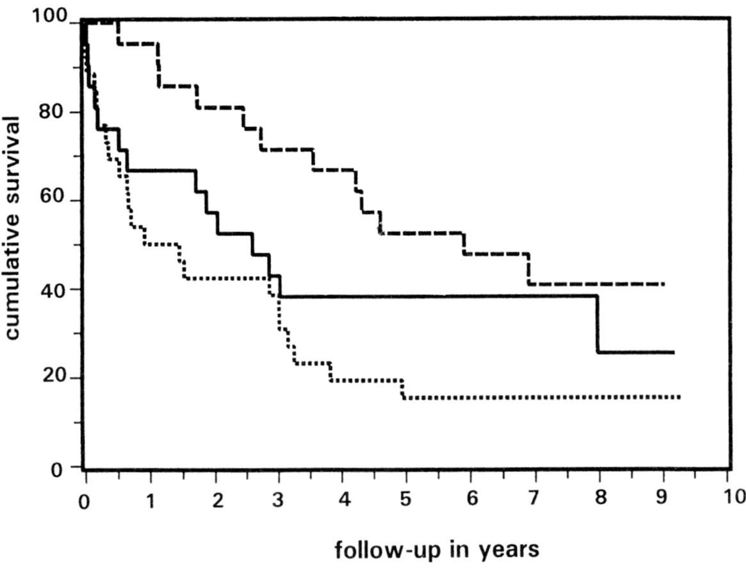

FIG. 1. Relationship between CD44v6 expression on primary colorectal carcinomas ($n = 68$) and percentage cumulative survival. The Kaplan–Meier curves show survival of patients with tumours with low (interrupted line), intermediate (solid line), and high (dotted line) CD44v6 expression. The differences are statistically significant at $P = 0.02$. Expression of CD44v6 was assessed immunohistochemically using the commercially available antibody VFF4 (Bender Med. Systems, Vienna).

of lung colonies (Sy et al 1991). It therefore appears that CD44 isoforms differ in the dissemination step they promote, and that for the spreading of cells from the site of injection into the draining lymph node, certain isoforms of CD44 work better than others. We therefore attempted to detect biochemical differences between CD44 isoforms and to unravel the normal functions of individual isoforms.

Biochemical properties of CD44 isoforms

In a survey of the biochemical properties of CD44s and CD44v, the molecules had several essentially similar characteristics, such as half life (our unpublished data). Some differences, however, were noticed. For example, in cell culture, CD44v4–v7 appears to be constitutively phosphorylated, while CD44s is not. These and other data have been used to construct a series of mutations aimed at defining the molecular properties CD44v proteins require in order to endue non-metastasizing cells with the ability to spread lymphogenically. Preliminary results suggest that the cytoplasmic tail and thus the interaction with the cytoskeleton of CD44v4–v7 are not prerequisites for mediating metastatic spreading.

Cross-linking studies also revealed differences between CD44s and CD44v proteins (our unpublished data). Experiments with metabolically labelled cells that express CD44v demonstrated that under non-reducing conditions and upon treatment of the cells with chemical cross-linking reagents, high molecular weight complexes could be immunoprecipitated with CD44-specific antibodies. Analysis of these complexes showed that they contained only CD44 variant isoforms; no other potential ligands or interacting proteins were detected. Furthermore, when CD44s and multiple different CD44v proteins were expressed in the same cell, only CD44v proteins were cross-linked into oligomers. Additionally, these oligomers contained only one type of CD44v, indicating homo-oligomerization even in the presence of other variants. In cells which express only CD44s, no high molecular weight complexes could be detected following cross-linking and immunoprecipitation with anti-CD44 antibodies. These data sugggest that the structural differences between CD44s and CD44v allow CD44v to form homo-oligomeric complexes on the cell surface, whereas CD44s cannot. These differences may contribute to an altered affinity of CD44v for ligands which in turn may increase the rate of metastasis.

Regulated affinity for hyaluronate

We have found that radiolabelled hyaluronic acid specifically binds to cells expressing CD44v, and that this binding is proportional to the levels of CD44v expressed on the cell surface. This is in contradiction to some reports suggesting that CD44v proteins do not bind to hyaluronate (Stamenkovic et al 1991), but this discrepancy can be explained by differences in cell type, culture conditions or the CD44 variant studied.

Preliminary evidence suggests that homomolecular association of CD44v may also increase its affinity for hyaluronic acid. Thus, when CD44v (but not CD44s) proteins are induced to cluster by antibodies, the affinity of CD44v (but not CD44s) for hyaluronate increases. An increase in affinity for hyaluronate may contribute to the metastatic process in several ways, all of which are speculative. Firstly, increased affinity for hyaluronate-containing tissues may allow cells to colonize these environments more efficiently. Conversely, a pericellular matrix of hyaluronate may decrease the affinity of a cell for surrounding tissues by interfering with adhesion processes, thus leading to enhanced detachment or simply increased mobility. Enhanced affinity to hyaluronate in a regulated manner, such as by the dynamic control of the oligomerization of CD44v proteins, could allow enhanced motility of cells along hyaluronate surfaces. Alternatively, enhanced hyaluronate binding by CD44v could facilitate the degradation of this matrix component, as it has been shown that CD44s is involved in the uptake and degradation of hyaluronic acid (Culty et al 1992). This could allow tumour cells to escape from entrapment within hyaluronate-rich environments. Moreover, enhanced hyaluronate binding by CD44v could act as a 'glue', facilitating binding of the variant part of the protein to ligands within the hyaluronate, e.g. cytokines which may enhance tumour cell survival and proliferation. Finally, what is true for hyaluronate may also be true for other ligands. Ordered presentation on the cell surface may thus enhance interaction with other perhaps yet unknown partners which may be limiting in metastasis.

A role for CD44 isoforms in the immune system

The highest turnover of hyaluronic acid in the adult can be attributed to the lymphatic system. Lymph nodes metabolize hyaluronic acid derived from peripheral tissues which they receive via the afferent lymph vessels (Fraser et al 1988). Many lymphocytes are able to follow the same pathway to enter the lymph nodes and the bulk of them are known to express high levels of CD44s on their surface. After activation of lymphocytes, a transient expression of CD44v-containing sequences encoded by exons v3 to v6 and perhaps additional v-exon sequences (Arch et al 1992, Koopman et al 1993, Mackay et al 1994) is observed. Most importantly, monoclonal antibodies which recognize an epitope encoded by CD44 exon v6 are able to inhibit activation *in vivo* of both B and T cells (Arch et al 1992). This strongly suggests that CD44v plays a crucial function during lymphocyte activation. In contrast, CD44s is permanently expressed on the surface of lymphocytes and its expression is elevated about twofold after antigenic stimulation. CD44 has therefore been used as a marker for memory or previously activated lymphocytes (Budd et al 1987). Unexpectedly, high levels of expression of CD44s and CD44v do not correlate with the capacity of lymphocytes to bind hyaluronate by polyclonal stimulation. Furthermore, upon ectopic expression of CD44(v4–v7) in lymphocytes, no hyaluronate binding capacity could be conferred, although the same cDNA was able to confer hyaluronate binding capacity when transfected into other cell types.

In order to study the role of CD44v in the immune system, we generated transgenic mice over-expressing rat CD44(v4–v7) permanently on the surface of their T cells (in collaboration with P. van der Putten, Basel). Resting T cells keep their CD44 molecules in an inactive conformation, at least as far as hyaluronate binding is concerned, and the protein expressed from the transgene is also inactive in this regard. Additionally, the distribution of T cells in the body is roughly normal. Activation of T cells may, however, change CD44 functionality and therefore their behaviour. Since activation requires CD44v expression (Arch et al 1992, Koopman et al 1993), the time needed for efficient surface expression should be identical or even shortened in the transgenic mice as compared with normal ones. In fact upon polyclonal and antigen-specific lymphocyte activators, the immune reponse is accelerated by approximately 24 hours, both *in vivo* and *in vitro* (in collaboration with M. Zöller, Heidelberg). Further detailed studies, including studies of repopulation kinetics upon adoptive transfer to irradiated host animals, are in progress.

Detailed study of the expression patterns of CD44v in normal embryonic and somatic tissues will yield important information concerning the normal function of CD44 variant proteins, particularly within migratory populations of cells. Similarly, further analysis of mice transgenic for CD44 variants, and the generation of mice in which CD44 variants are 'knocked out', will reveal how CD44v proteins function in the whole animal. This information will be of direct relevance to the role of CD44v in the metastatic phenotype and will help us to design experiments to determine how the normal function of CD44v is subverted into the development of metastasis.

Acknowledgements

We thank I. Kammerer and D. Nichol for their help in the preparation of the manuscript. This work was supported by a grant from the Deutsche Forschungsgemeinschaft (He 551/8-1).

References

Arch R, Wirth K, Hofmann M et al 1992 Participation in normal immune responses of a splice variant of CD44 that encodes a metastasis-inducing domain. Science 257:682–685

Aruffo A, Stamenkovic I, Melnick M, Underhill CB, Seed B 1990 CD44 is the principal cell surface receptor for hyaluronate. Cell 61:1303–1313

Budd RC, Cerottini J-C, MacDonald HR 1987 Phenotypic identification of memory cytolytic T lymphocytes in a subset of LYT-2^+ cells. J Immunol 138:1009–1013

Camp RL, Kraus TA, Puré E 1992 Variations in the cytoskeletal interaction and posttranslational modification of the CD44 homing receptor in macrophages. J Cell Biol 115:1283–1292

Culty M, Miyake K, Kincade PW, Silorski E, Butcher EC, Underhill CB 1990 The hyaluronate receptor is a member of the CD44 (H-CAM) family of cell surface glycoproteins. J Cell Biol 111:2765–2774

Culty M, Nguyen HA, Underhill CB 1992 The hyaluronan receptor (CD44) participates in the uptake and degradation of hyaluronan. J Cell Biol 116:1055–1062

Dall P, Heider K-H, von Minckwitz G, Kaufmann M, Ponta H, Herrlich P 1994 Surface protein expression and mRNA-splicing analysis of CD44 in uterine cervical cancer and normal cervical epithelium. Cancer Res 54:3337–3341

Favrot MC, Combaret V, Lasset C, Bérard CL 1993 CD44—a new prognostic marker for neuroblastoma. N Engl J Med 329:1965

Finn L, Dougherty G, Finley G, Meisler A, Becich M, Cooper DL 1994 Alternative splicing of CD44 pre-mRNA in human colorectal tumors. Biochem Biophys Res Commun 200:1015–1022

Fox SB, Gatter KC, Jackson DG et al 1993 CD44 and cancer screening. Lancet 342:548–549

Fraser JR, Kimpton WG, Laurent TC, Cahill RNP, Vakakis N 1988 Uptake and degradation of hyaluronan in lymphatic tissue. Biochem J 256:153–158

Günthert U, Hofmann M, Rudy W et al 1991 A new variant of glycoprotein CD44 confers metastatic potential to rat carcinoma cells. Cell 65:13–24

Heider K-H, Hofmann M, Horst E et al 1993a A human homologue of the rat metastasis-associated variant of CD44 is expressed in colorectal carcinomas and adenomatous polyps. J Cell Biol 120:227–233

Heider K-H, Dämmrich J, Skroch-Angel P et al 1993b Differential expression of CD44 splice variants in intestinal and diffuse type human gastric carcinoma and normal gastric mucosa. Cancer Res 53:4197–4203

Herrlich P, Rudy P, Hofmann M et al 1993 CD44 and splice variants of CD44 in normal differentiation and tumour progression. In: Hemler ME, Mihich E (eds) Cell adhesion molecules. Plenum, New York, p 265–288

Hynes RO 1992 Integrins: versatility, modulation, and signaling in cell adhesion. Cell 69:11–25

Kaufmann M, Heider K-H, Sinn H-P, von Minckwitz G, Ponta H, Herrlich P 1994 Surface expression of distinct CD44 variant exon epitopes on primary breast cancer strongly correlates with poor survival. N Engl J Med, in press

Koopman G, Heider K-H, Horst E et al 1993 Activated human lymphocytes and aggressive non-Hodgkin lymphomas express a homologue of the rat metastasis-associated variant of CD44. J Exp Med 177:897–904

Lesley J, Hyman R, Kincade PW 1993 CD44 and its interaction with the cellular matrix. Adv Immunol 54:271–335

Li H, Hamou M-F, de Tribolet N et al 1993 Variant CD44 adhesion molecules are expressed in human brain metastases but not in glioblastomas. Cancer Res 53:5345–5349

Mackay CR, Terpe H-J, Stauder R, Marston WL, Stark H, Günthert U 1994 Expression and modulation of CD44 variant isoforms in humans. J Cell Biol 124:71–82

Matsumara Y, Tarin D 1992 Significance of CD44 gene products for cancer diagnosis and disease evaluation. Lancet 340:1053–1058

Mayer B, Jauch KW, Günthert U et al 1993 De-novo expression of CD44 and survival in gastric cancer. Lancet 342:1019–1022

Reber S, Matzku S, Günthert U, Ponta H, Herrlich P, Zöller M 1990 Retardation of metastatic tumor growth after immunization with metastasis-specific monoclonal antibodies. Int J Cancer 46:919–927

Rudy W, Hofmann M, Schwartz-Albiez R et al 1993 The two major CD44 proteins expressed on a metastatic rat tumor cell line are derived from different splice variants: each one individually suffices to confer metastatic behavior. Cancer Res 53:1262–1268

Salmi M, Grön-Virta K, Sointu P, Grenman R, Kalimo H, Jalkanen S 1993 Regulated expression of exon v6 containing isoforms of CD44 in man: downregulation during malignant transformation of tumors of squamocellular origin. J Cell Biol 122:431–442
Screaton GR, Bell MV, Jackson DG, Cornelis FB, Gerth U, Bell JI 1992 Genomic structure of DNA encoding the lymphocyte homing receptor CD44 reveals at least 12 alternatively spliced exons. Proc Natl Acad Sci USA 89:12160–12164
Seiter S, Arch R, Komitowski D et al 1993 Prevention of tumor metastasis formation by anti-variant CD44. J Exp Med 177:443–455
Stamenkovich I, Aruffo A, Amiot M, Seed B 1991 The hematopoietic and epithelial forms of CD44 are distinct polypeptides with different adhesion potentials for hyaluronate-bearing cells. EMBO (Eur Mol Biol Organ) J 10:343–348
Sy MS, Guo Y-J, Stamenkovic I 1991 Distinct effects of two CD44 isoforms on tumor growth in vivo. J Exp Med 174:859–866
Tölg C, Hofmann M, Herrlich P, Ponta H 1993 Splicing choice from ten variant exons establishes CD44 variability. Nucleic Acids Res 21:1225–1229
van Weering DHJ, Baas PD, Bos JL 1994 A PCR-based method for the analysis of human CD44 splice products. PCR Methods Appl 3:100–106
Wheatley SC, Isacke CM, Crossley PH 1993 Restricted expression of the hyaluronan receptor, CD44, during postimplantation mouse embryogenesis suggests key roles in tissue formation and patterning. Development 199:295–306
Wielenga VJM, Heider K-H, Offerhaus GJA et al 1993 Expression of CD44 variant proteins in human colorectal cancer is related to tumor progression. Cancer Res 53:4754–4756
Wirth K, Arch R, Somasundaram C et al 1993 Expression of CD44 isoforms carrying metastasis-associated sequences in newborn and adult rats. Eur J Cancer 29A:1177–1183

DISCUSSION

Labow: Like myself, wild-type CD44 is short and stubby! Could the increased metastatic potential of the variant simply be because it is longer and, therefore, it is easier for it to interact with a charged molecule such as hyaluronic acid?

Herrlich: If that were the case, why would 10 different exons with no sequence similarity be needed to accomplish this? Why wouldn't the cells produce extended versions of CD44 by simply inserting a repeat of one and the same amino acid sequence? We prefer the alternative interpretation that the additional exon sequences either modulate (e.g. up-regulate) adhesion to hyaluronate, or recruit additional ligands.

Wagner: You mentioned that the glycosylation of the different exons is also important for the function of CD44, but you didn't say anything specific.

Herrlich: This is because we are not sure which type of glycosylation is essential for its function, for example, in promoting metastasis. The cells that bind to hyaluronic acid produce very large glycosylated CD44 molecules with carbohydrate components three to four times the molecular weight of the protein. In certain cells that do not bind hyaluronic acid, there appears to be less receptor glycosylation. Consequently, we think that the glycosylation contributes to the binding, but because we have not identified exactly which parts of the structure are essential, we can't prove this yet.

Ruggeri: What can one conclude from the last two papers in terms of a general mechanism for metastasis? It intrigues me that metastatis requires loss of adhesive function and then reaquisition of adhesion at a distant site.

Herrlich: Imagine you were standing in a crowd of densely packed people and you wished to move: you would definitely need to get rid of your adhesion to the crowd. This is the concept Walter Birchmeier has been trying to get across (Birchmeier et al 1995, this volume). On the other hand, once you are moving, some adhesion is beneficial: remember when you last tried to walk on an icy surface. You would have been able to move more easily had you been able to adhere. I believe this analogy is applicable to metastasis. A cell that cannot adhere is probably doomed to die.

Loss of E-cadherin is achieved by one of several mechanisms, ranging from deletion to post-translational inactivating modification. Gain of adhesive ability is achieved by alternatively splicing the CD44 nuclear transcript. Many laboratories have detected alternatively spliced CD44 transcripts by PCR in a number of human tumours. Our own laboratory has concentrated mainly on detecting the new epitopes on the cell surface.

Ruggeri: These are very fine, very interesting observations, but it sounds like they're sort of restricted to specific situations. How does it eventually come together?

Herrlich: What is very interesting is that several of these molecules, like DCC (encoded by Fearon & Vogelstein's gene 'defective in colon cancer') and E-cadherin are expressed in a pattern that is the mirror image to that of CD44. In neural development, neural crest cells never carry CD44 when they have DCC and E-cadherin on them, but when they start migrating they lose E-cadherin and only regain it later on. In the meantime, they express CD44 variants. I have no idea how this is regulated, but it seems like mutual, e.g. discordant, regulation.

Birchmeier: Another interesting point is that the pancreatic carcinoma cell line that you used for transfection with a CD44 variant is already a fully differentiated line. Apparently, loss of E-cadherin alone doesn't make it fully metastatic. But when you look for expression of CD44 variants in the progression of colon carcinomas, you often see the variants *in vivo* appearing relatively early when the cells are still differentiated. So, in some cases the disappearance and reappearance of 'metastasis' molecules do not happen in a fixed order.

Hynes: They clearly don't in the cascade that one sees presented all the time for the development of colon carcinoma. There's a preferential order to these events, but it's not strict. Another thing that's worth keeping in mind is that very often one is dealing with a small population that then takes the next step. You are selecting for the metastatic variants from among a large population of varying cells—for instance, once p53 is mutated, the rate of other mutations increases, and so the number of variants that appear in the pool from which you can select a more metastatic variant goes up. You shouldn't think of the whole tumour as progressing along a rigid pathway.

Wagner: Since antibodies to the variant regions of CD44 are inhibitory, would it work if you injected fragments of hyaluronic acid into the blood?

Herrlich: This has not yet been done. I wonder whether it would be effective.

Mimicking the work with the CS-1 peptide (e.g. Elices 1995, this volume), we have attempted to suppress metastases by injecting peptides corresponding to variant exons. We have not yet been successful in suppressing metastatic functions, probably because the half-life was too short or the concentration was not sufficient.

Birchmeier: That might be because you didn't have the right sugars on them.

Hynes: You stressed the hyaluronate binding activity of CD44. There are various reports of its binding to collagen and fibronectin. Is any of this affected by these alternative splices?

Herrlich: When examining the tumour cell lines that either express CD44s or CD44s plus variants, we have not seen much of a difference between collagen and fibronectin binding. These studies need to be repeated with the soluble isoforms. A soluble isoform can be cross-linked by a second antibody; when we do this, we elevate the binding to hyaluronic acid. This should also be done with collagen and fibronectin.

Wagner: It sounds like some of the variants may be mucin-like molecules: could they present sugars to the selectins?

Herrlich: A specific glycosylation requiring a specific fucosyl transferase has recently been linked to a subpopulation of CD44 molecules (J. Le Pendu, unpublished results). In collaboration with Jacques Le Pendu we hope to find out whether the modification has any functional significance for the metastatic process.

Hogg: A speculative point along the same lines: perhaps it is only important to have many domains in CD44 in order to carry these large charged sugars. Maybe the desired effect is to create an anti-adhesive molecule that would help cells move from the primary tumour to somewhere else.

Herrlich: We have measured where the bulk of the sugar is attached to; this seems to be in the constant, N-terminal region. The variant exons seem to be under-glycosylated. This is done simply by cutting the molecule into pieces and looking at the molecular weight of the fragments. From this result we tentatively conclude that it is not de-adhesion that matters, but a gain of function. Also to this end, the inhibition of metastasis by antibody speaks against de-adhesion.

Etzioni: Regarding Burkitt's lymphoma, did you look at both the African and the American type? The American is EBV-negative. We know from the clinical data that Burkitt's lymphoma is the fastest growing lymphoma. You don't have a variant of CD44 in this lymphoma, so what is its importance?

Herrlich: We have looked at Stephen Pals' collection of Burkitt's lymphoma, which is small. We are now collaborating with Henri Jacques Delecluse in Lyon, France, to address the question you raise. One possibility is that aggressive disseminating lymphomas escape from repression and synthesize CD44v6 exon variants.

Haskard: CD44 has also been referred to as the Hermes antigen and has been implicated in lymphocyte adhesion to endothelial cells. Have you got any evidence from antibody inhibition studies as to whether blocking CD44 *in vivo* alters lymphocyte traffic in any way?

Herrlich: In transgenic mice with targeted expression of CD44v4–v7 to T cells, the distribution of lymphocytes and their subsets doesn't seem to be changed at all. So, at least in an equilibrium state, there is apparently no change in their trafficking. Maybe this is to be expected. Recall that I showed that there was an absence of affinity to hyaluronate in CD44v4–v7-expressing T cells, but good binding upon expression of the same molecule on tumour cells. Regulated affinity may also affect the interaction with other ligands. If the majority of cells remained unstimulated, the transgene product would not affect trafficking. However, if we irradiate animals and then transfer bone marrow from the transgenic animals into them, repopulation is faster, as if the transgenic molecule on the cell surface confers a homing or a growth advantage.

Haskard: But have you tried any antibody inhibition experiments with anti-CD44 to see whether you can inhibit lymphocyte migration, either to lymph nodes or to inflammatory lesions?

Herrlich: In vitro, when we explore the difference between the transgenic population and the non-transgenic control population in the mixed lymphocyte reaction and find that the transgenic lymphocytes react faster, we can reverse the acceleration of entry into S phase by using an antibody to the exon v6 epitope of CD44. This suggests that there is a critical CD44 interaction that takes place in a mixed lymphocyte culture during *in vitro* activation. Preliminary results have even suggested that soluble CD44–IgG fusion proteins can block the mixed lymphocyte reaction, suggesting that there is a specific competitive interaction in the population of spleen or lymph node cells.

Haskard: The question I'm really getting at is whether you think CD44 has any role in lymphocyte–endothelial cell interactions.

Herrlich: We have no evidence for this.

Pober: Bart Haynes initially reported that CD44 participates in CD2-mediated rosetting with LFA-3 on erythrocytes (Hale et al 1989) and several groups have found that CD44 on human T cells is a co-stimulatory signal apparently interacting with the CD2 pathway (Shimizu et al 1989, Huet et al 1989, Denning et al 1990, Conrad et al 1992). This co-stimulatory function is consistent with the notion that the cross-linked molecule might give an advantage to T cells.

Regarding CD45, the B exon is present on activated T blasts and it's also present on resting naive T cells, but the expressed CD45 isoform that includes that exon is probably different between those two subsets. So if you simply carry out fluorescence-activated cell sorting (FACS) and look for the expression of the B exon-encoded epitope, that will tell you which CD45 isoform is actually up-regulated. Have you done biochemistry to see whether this is the full CD45 (containing ABC) or whether it's just CD45 B?

Herrlich: Stephen Pals analysed human tonsil T cells by FACS and found elevated CD44 exon v6 epitope expression on a population that carries CD45RA and CD45RO simultaneously. This population represents recently activated T cells.

Pals: If you look in human tonsil T lymphocytes, then you find that naive T cells express virtually no variants. The CD45RO memory T cells show a slight increase in CD44 variant expression, but the level is still extremely low. However, in the CD45RA RO double-positive cells in tonsil, in about 2% of the total T cell population that is undergoing conversion from the naive to the memory phenotype, CD44 variants are strongly expressed (Griffioen et al 1994).

Pober: In these experiments you are taking the expression of B epitope as evidence for partial T cell activation. It is important to know which molecule actually carries that epitope, because expression of the full ABC isoform of CD45 would not be evidence for activation, whereas the isoform that contains the B epitope but has the A epitope spliced out would be.

Herrlich: We have only looked at the A epitope using antibodies; this did not show an enhancement. So it is probably an outspliced form similar to the human CD45 forms mentioned above.

The cross-linking of CD44 that causes a co-stimulatory signal in T cells may be different from that which we see as having a metastasis-promoting effect in the tumour cells, because in the tumour cells the C-terminal end seems redundant. Perhaps we are looking at an advantage created by an adhesion process that does not seem to depend on the cytoplasmic tail. However, the T cells, where we see stimulation and perhaps even a faster stimulation when there is expression of these variants, may use a signalling pathway dependent on the CD44 tail.

Pober: The antibody co-stimulation is confusing though, because the data indicate that signalling through CD44 is probably not a direct effect: it depends upon signalling through CD2. The nature of the interaction between CD44 and CD2 is not clear.

Sanchez-Madrid: Have you found any CD44 variants in T cells that are activated *in vivo* by inflammatory conditions?

Herrlich: Among our collaborators, Stephen Pals has done influenza-specific stimulation and Margot Zöller has used TNP hapten-specific activation *in vivo*, all of which have resulted in the transient appearance of CD44 splice variants. So there is at least indirect evidence that this is also occurring *in vivo*.

Hogg: Are the same exons expressed by activated T cells in all individuals?

Herrlich: We don't know.

Riethmüller: Have you looked at the *in vivo* migration pattern in these transgenic T cells?

Herrlich: The homing pattern in an adoptive transfer of bone marrow to irradiated recipients is accelerated.

References

Birchmeier W, Hülsken J, Behrens J 1995 E-cadherin as an invasion suppressor. In: Cell adhesion and human disease. Wiley, Chichester (Ciba Found Symp 189) p 124–141

Conrad P, Rothman BL, Kelley KA, Blue ML 1992 Mechanism of peripheral T cell activation by coengagement of CD44 and CD2. J Immunol 149:1833–1839

Denning SM, Lee PT, Singer KH, Haynes BF 1990 Antibodies against the CD44 p80 lymphocyte homing receptor molecule augment human peripheral blood T cell activation. J Immunol 144:7–15

Elices MJ 1995 The integrin $\alpha_4\beta_1$ (VLA-4) as a therapeutic target. In: Cell adhesion and human disease. Wiley, Chichester (Ciba Found Symp 189) p 79–90

Griffioen AW, Horst E, Heider KH et al 1994 Expression of CD44 splice variants during lymphocyte activation and tumor progression. Cell Adhes Commun 2:195–200

Hale LP, Singer KH, Haynes BF 1989 CD44 antibody against In(lu)-related p80 lymphocyte homing receptor molecule inhibits the binding of human erythrocytes to T cells. J Immunol 143:3944–3948

Huet S, Groux H, Caillou B, Valentin H, Prieur AM, Bernard A 1989 CD44 contributes to T cell activation. J Immunol 143:798–801

Shimizu Y, Van Seventer GA, Siraganian R, Wahl L, Shaw S 1989 Dual role of the CD44 molecule in T cell adhesion and activation. J Immunol 143:2457–2463

Early metastasis of human solid tumours: expression of cell adhesion molecules

K. Pantel, G. Schlimok*, M. Angstwurm, B. Passlick†, J. R. Izbicki†, J. P. Johnson and G. Riethmüller

*Institut für Immunologie, Ludwig-Maximilians-Universität, Goethestrasse 31, 80336 München, *Medizinische Klinik II, Zentralklinkum Stenglinstrasse 2, 86156 Augsburg and †Abt. für Allgemeinchirurgie, Universitätskrankenhaus Eppendorf, Martinistrasse 52, 20251 Hamburg, Germany*

Abstract. Loss and gain of cell surface molecules determines the mobilization, emigration and invasiveness of epithelial cancer cells. As a first approach to gain further insight into these processes, we have followed two strategies: (1) to identify tumour cells which have disseminated early from primary carcinomas and to obtain information about the phenotype and prognostic significance of these cells; and (2) to identify molecular changes occurring in primary tumour cells at the time they develop their metastatic potential. Our analyses indicate that changes in the adhesive properties of solid tumour cells, such as down-regulation of desmosomal proteins (e.g. plakoglobin) and neo-expression of ICAM-1 or MUC18, are important determinants of the metastatic capability of individual malignant cells. The expression pattern of these cell adhesion molecules during tumour progression appears to reflect a disturbance at the level of the molecular elements normally responsible for controlling their expression. The outlined current strategies for detection, characterization and antibody therapy of cancer micrometastasis can be applied to the secondary prevention of metastatic disease in patients with minimal residual cancer.

1995 Cell adhesion and human disease. Wiley, Chichester (Ciba Foundation Symposium 189) p 157–173

It is an old truism that metastasis accounts for the majority of cancer deaths in industrialized countries. In these societies, the most frequent cancers are derived from epithelia of the gastrointestinal and urogenitary tract, as well as of mammary ducts and bronchi. Although these tumours are being diagnosed at increasingly earlier stages, allowing for a high rate of local cure, mortality has not substantially decreased—essentially because patients continue to succumb to metastatic disease. This clearly points to dissemination of tumour cells at an early stage of tumour growth, and implies that the acquisition of at least some characteristics of metastatic behaviour can occur prior to attainment of the unrestrained growth observed in fully developed tumours. In this context, it is important to consider that tumorigenesis and metastasis development are not necessarily the result of the same genetic changes (Fidler & Radinsky 1990, Liotta et al 1991). In fact, defined mutations in proto-oncogenes often confer tumorigenic characteristics to cells while their metastatic potential is not activated (Price et al 1989). This is not surprising in light of the fact that the characteristics of these two processes can, at least in part, be separated. Development of unrestricted growth is focused on the acquisition of growth factor independence and loss of contact inhibition while development of metastasis is focused on enhanced cell motility, expression of proteases and changes in the expression of cell and matrix adhesion molecules (Hart et al 1989).

As a first approach to gain more insight into the determinants of metastatic potential in solid human tumours, we have followed two strategies: (1) to identify, at as early a stage as possible, tumour cells which have disseminated from primary carcinomas and to obtain information about the phenotype and prognostic significance of these cells; and (2) to identify molecular changes occurring in primary tumour cells at the time when they develop metastatic potential.

The early seeding of disseminated tumour cells, or micrometastasis, can be observed using immunocytochemical techniques and antibodies which distinguish between the tumour cell and the microenvironment under examination. Monoclonal antibodies to cytokeratin components which are restricted to simple epithelia (e.g. cytokeratin 18 [CK18]) have proven to be sensitive and specific probes for the detection of individual micrometastatic cells in the bone marrow of patients with various epithelial tumours (Riethmüller & Johnson 1992, Pantel et al 1993a). The incidence of these cells in patients without any signs of overt metastasis (stage M_0) ranges from about 15% to more than 30% (Table 1). For the detection of epithelial cells in bone marrow, monoclonal antibodies to cytokeratins are clearly

TABLE 1 Incidence of CK18-positive cells in bone marrow of patients with epithelial cancer with no clinical signs of metastasis (M_0)

Tumour origin	*Number of patients*	*Number of patients with CK18-positive cells (%)*
Breast	116	35 (30.2)
Lung (NSCLC)	101	26 (25.7)
Stomach	80	25 (31.2)
Colon/rectum	195	53 (27.2)
Prostate	80	27 (33.8)
Kidney	76	12 (15.8)
Bladder	66	15 (22.7)
Control group[a]	215	6 (2.8)

[a]Patients without evidence of an epithelial malignancy.
NSCLC, non-small-cell lung carcinoma.

superior to monoclonal antibodies against mucin-like cell membrane proteins, which cross-react with haemopoietic cells (Schlimok et al 1991, Delsol et al 1984). Morphological criteria used by some authors for differentiation of false-positive cells stained by epithelial membrane antibodies (Cote et al 1991, Diel et al 1992, Brugger et al 1994) are rather unreliable because of the marked morphological heterogeneity of aspirated and processed carcinoma cells. Previous investigations suggesting ectopic expression of CK18 in mesenchymal cells of lymph nodes (Jahn et al 1987, Järvinen et al 1990) were not confirmed on bone marrow by double marker analyses (Schlimok et al 1987, Pantel et al 1994a). In the case of patients with prostatic cancer, the histogenetic origin of CK18-positive cells in bone marrow was directly proven by coexpression of prostate-specific antigen on these cells (Riesenberg et al 1993). However, it is unclear whether, in the very few control patients with no malignant disease presenting CK-positive cells in marrow specimens, these cells are normal epithelial cells (Table 1) or malignant cells disseminated from a small primary carcinoma undetected at the time of bone marrow sampling. The latter conclusion is supported by the subsequent diagnosis of colon cancer in one control patient, who displayed CK18-positive cells in a previous bone marrow analysis (Pantel et al 1993b).

In patients with mammary, lung and gastrointestinal carcinomas, the presence of CK18-positive cells in bone marrow was found to be associated both with the occurrence of manifest metastases and a reduced relapse-free survival time, supporting the clinical relevance of individual disseminated tumour cells (Schlimok et al 1991, 1992, Lindemann et al 1992, Pantel et al 1993c). The biology of micrometastatic cells remains, however, poorly understood. This ignorance is particularly disturbing in diseases like colorectal cancer in which the frequent finding of bone marrow micrometastasis revealed by immunocytochemistry has proved to be a strong and independent predictor of overall metastatic relapse, though manifest skeleton metastasis occurs rarely (Lindemann et al 1992). Thus, tumour cells detected in bone marrow at the time of operation on the primary tumour may not necessarily have the potential to form clinically detectable metastases within the remaining lifespan of the patient; this may be determined by the more aggressively growing liver metastases, while the cells in bone marrow may remain dormant for years (Pantel et al 1993b).

The extremely low frequency of micrometastatic cells greatly hampers attempts to obtain more specific information on their biological properties (Pantel 1992, 1994a). First attempts have been made to phenotype these cells by immunocytochemical double-marker analysis (Table 2). These studies indicate that individual carcinoma cells disseminated to bone marrow represent a selected population of cancer cells which, however, still express a considerable degree of heterogeneity. These cells rarely express the nuclear proliferation-associated antigens Ki-67 and p120, indicating that they are resting in G0 or early G1 phase

TABLE 2 Phenotype of cytokeratin-positive epithelial tumour cells in bone marrow

Marker	*Tumour origin*	*No. of patients with marker-positive/ cytokeratin-positive cells*	*Reference*
Histogenetic marker proteins			
PSA	Prostate	5/13 (38.5%)	Riesenberg et al 1993
CD45	Breast/colorectum	0/18	Pantel et al 1994a, Schlimok et al 1987
Vimentin	Breast/colorectum	0/13	Pantel et al 1994a
Growth factor receptors			
EGF-R	Breast	10/37 (27.0%)	Schlimok & Riethmüller 1990
	Colorectum	4/15 (26.7%)	Schlimok & Riethmüller 1990
erbB2	Breast	48/71 (67.6%)	Pantel et al 1993b
	Colorectum/stomach	14/50 (28.0%)	Pantel et al 1993b
Transferrin-R	Breast	17/59 (28.8%)	Schlimok & Riethmüller 1990
	Colorectum	7/17 (41.1%)	Schlimok & Riethmüller 1990
p53 tumour suppressor protein			
	Colorectum/stomach	4/45 (8.9%)	Pantel et al 1993d
	Lung, NSCLC	0/13	Pantel et al 1993d
MHC class I antigens			
	Breast	9/26 (34.6%)	Pantel et al 1991
	Colon/stomach	37/65 (56.9%)	Pantel et al 1991
Adhesion molecules			
ICAM-1	Lung, NSCLC	13/31 (41.9%)	Pantel et al 1992
Plakoglobin	Lung, NSCLC	4/12 (33.3%)	K. Pantel, unpublished results
	Colorectum/stomach	4/13 (30.8%)	K. Pantel, unpublished results
17-1A	Breast	7/16 (43.8%)	K. Pantel, unpublished results
	Colorectum	4/6 (66.7%)	K. Pantel, unpublished results
Proliferation-associated antigens			
Ki-67	Breast	1/12 (8.3%)	Pantel et al 1993b
	Colorectum/stomach	0/21	Pantel et al 1993b
p120	Breast	1/11 (9.1%)	Pantel et al 1993b
	Colorectum/stomach	9/32 (28.1%)	Pantel et al 1993b

NSCLC, non-small-cell lung carcinoma.

of the cell cycle, which is consistent with the phenomenon of tumour cell dormancy (Pantel et al 1993b). Nevertheless, a considerable fraction of them overexpress the tyrosine kinase receptor erbB2 (Pantel et al 1993b) and exhibit a proliferative response in cell cultures (Pantel et al 1994b), which suggests that they possess an increased growth potential *in situ*.

Among the numerous factors affecting the metastatic capability of malignant cells encased in a solid tumour, one of the most important determinants is a change in their adhesive properties. Loss of homotypic adhesion is one of the first steps required for the successful dissemination of tumour cells (Hart et al 1989). In the case of epithelial organs, a network of intercellular adhesive junctions is responsible for the tight integration of an individual cell within a particular tissue (Schwarz et al 1990). The adherens junction complex is organized around a transmembrane cadherin protein that organizes a complex of cytoplasmatic proteins, including α-catenin, β-catenin and plakoglobin, a β-catenin relative found in desmosomes (Schwarz et al 1990). The adherin–catenin complex mediates adhesion, cytoskeletal anchoring and signalling. Catenins can also form a complex with the product of the tumour suppressor gene *APC*, which may mediate transmission of a growth regulatory signal (Peifer 1993). Using our double-marker technique, we studied expression of plakoglobin on isolated micrometastatic carcinoma cells in bone marrow. Our results demonstrate that the expression of plakoglobin is frequently reduced or absent in these cells (Table 2), suggesting that down-regulation of its expression might be an important determinant of the disseminative capacity of an epithelial tumour cell. It would now be of great interest to evaluate whether the biological effects of plakoglobin are mediated by regulation of cell adhesion via the APC–catenin complex.

Another interesting homophilic cell–cell adhesion molecule is the epithelial glycoprotein 40 (EGP40, also known as GA-733-2, ESA, KSA and the 17-1A antigen) encoded by the GA-733-2 gene, which is expressed on the basolateral cell surface in most human simple epithelia (Göttlinger et al 1986). The protein is also expressed in the vast majority of carcinomas and has attracted attention as a tumour marker (Göttlinger et al 1986). Recently, Litvinov et al (1994) provided first evidence that EGP40 is an epithelium-specific intercellular adhesion molecule, which acts in a calcium-independent manner. In its biological behaviour, EGP40 resembles members of the immunoglobulin superfamily, which includes cell adhesion molecules. Therefore the name Ep-CAM was proposed. Our present double-marker analysis demonstrates the frequent absence of Ep-CAM or 17-1A antigen on isolated breast cancer cells disseminated to bone marrow (Table 2), which supports a role of this protein in early dissemination of epithelial tumour cells.

Nevertheless, four out of six patients with colorectal cancer exhibited coexpression of 17-1A on CK18-positive cells in marrow (Table 2), suggesting that this antigen might be an appropriate target for passive immunotherapy with monoclonal antibodies in this type of cancer. To test this hypothesis, we treated colorectal cancer patients with no distant metastases (stage Dukes C) in a randomized clinical trial with the murine monoclonal antibody 17-1A against the 17-1A antigen (Riethmüller et al 1994). The antibody mediates antibody-dependent cellular cytotoxicity of colorectal carcinoma lines *in vitro* and suppresses the growth of human tumours grafted into nude mice (Herlyn et al 1980). Patients in this adjuvant trial received a total of 900 mg of the 17-1A monoclonal antibody over 20 weeks, starting 2 to 3 weeks after primary surgery. Following a median follow-up of 5 years, therapy with antibody was found to reduce the overall death rate by 30% and decrease the recurrence rate by 27% (Riethmüller et al 1994). These data contrast with the results of numerous trials with 17-1A antibody in advanced tumours where anecdotal remissions were observed in only a few patients and no benefit for survival could be secured (Riethmüller et al 1993). Thus, by carefully selecting the disease stage at which therapy is initiated, the efficacy of antibody therapy is comparable with that of other adjuvant therapies (Moertel et al 1991, Krook et al 1991), while the toxicity of 17-1A treatment is considerably lower (Riethmüller et al 1994).

Single tumour cells present in an easily accessible environment such as bone marrow should be preferential targets for immune effector cells, although it is known that micrometastatic cells can survive for extended periods of time without being killed by the immune system (Riethmüller & Johnson 1992). This discrepancy might be due to non-recognition of these cells by immune effector cells and/or to an anergic state of the effector cells. In a first attempt to address the problem of tumour cell recognition, micrometastatic cells were typed for the expression of MHC molecules which present tumour-specific peptides via the T cell receptor to T lymphocytes (Hämmerling et al 1989). A down-regulation of HLA class I molecules had been previously observed in several different primary carcinomas and had in some cases been correlated with poor prognosis (Tanaka et al 1988). Disseminated HLA class I-negative tumour cells may preferentially generate metastases, perhaps in part by escaping from recognition by cytotoxic T lymphocytes. Thus, the frequent reduction in HLA expression observed in breast carcinoma patients (Table 3) is consistent with the high incidence of overt skeletal metastasis occurring in these tumours. The overall incidence of HLA-negative tumour cells in the bone marrow is higher than has been reported for the same tumours examined as primary lesions (Hämmerling et al 1989), an observation which supports the assumption that these cells do indeed have a selective survival advantage.

The interaction of immune cells with their targets is also influenced by additional molecules on the target cells, in particular by the cell adhesion molecules which interact with leukocyte ligands (Springer 1990). One of these molecules, the intercellular adhesion molecule 1 (ICAM-1), mediates leukocyte binding through its interaction with the integrins $\alpha_L\beta_2$ (LFA-1) and $\alpha_M\beta_2$ (Mac-1). ICAM-1 is not normally expressed by epithelial cells but can be induced

TABLE 3 Deficient expression of HLA class I molecules on individual disseminated carcinoma cells in bone marrow

Groups	*Number of patients*	*Number of patients with HLA class I-negative CK18-positive cells in marrow*[a] *(% of total)*
Origin of primary tumour		
Breast	26	17 (65.4%)
Colon	17	5 (29.4%)
Stomach	11	3 (27.3%)
Differentiation grade of primary tumour		
Grade 2	23	7 (30.4%)[b]
Grade 3	21	13 (61.9%)

[a]Determined using antibodies CK2 and W6/32 in double-labelling procedure (Pantel et al 1991).
[b]Difference between differentiation grades is significant ($P<0.05$; χ^2-test).

by a variety of stimuli and is sometimes expressed by carcinoma cells *in vivo* (Passlick et al 1994). Expression of ICAM-1 on CK18-positive cells in the bone marrow of patients with completely resectable non-small-cell lung carcinoma (NSCLC), appeared to be of prognostic significance. As can be seen from Table 4, those patients with ICAM-1-positive tumour cells had a longer disease-free period and an increased survival time when compared to those patients with ICAM-1-negative tumour cells. A similar prognostic significance of ICAM-1 expression on primary tumour cells of NSCLC could not be ascertained (Passlick et al 1994); this observation is consistent with studies on renal cell carcinoma in which the expression of ICAM-1 by the primary tumour was correlated with well differentiated tumours having a good prognosis (Tomita et al 1990).

The current strategies outlined here for detection, characterization and antibody therapy of cancer micrometastasis can be applied for secondary prevention of metastatic disease in patients with operable primary carcinomas. The remarkable specificity and sensitivity of immunocytochemical cytokeratin assays for micrometastasis detection supports their development into a monitoring device for adjuvant therapies. Quantitative aspects (e.g. number of micrometastatic cells per volume of marrow aspirate) as well as qualitative ones (e.g. expression of oncogenes) may have to be considered before this assay is introduced into the current practice of tumour staging.

In some human tumours, the development of metastatic potential can be strongly correlated with particular characteristics of the primary tumour. In cutaneous melanoma, studies by Clark et al (1989) indicate that cells with the capacity to produce metastasis first separate from the primary lesion when the tumour enters the vertical growth phase. As tumours generally remain thin during radial growth phase, this observation may explain why the vertical

TABLE 4 ICAM-1 expression on individual disseminated carcinoma cells in bone marrow of patients with operable non-small-cell lung carcinoma (stage M_0)

Groups	*No. patients per group*	*ICAM-1 expression on CK18-positive cells in bone marrow*[a]	
		Positive	*Negative*
Total	31	13 (41.9%)	18 (58.9%)
Tumour histology			
adenomatous	14	3	11
squamous	13	8[b]	5
miscellaneous	4	2	2
ICAM-1 expression on primary tumour			
positive	10	6	4
negative	15	3[c]	12
Follow-up[d]			
Local relapse	6	0	6
Metastasis	8	2	6
Death	5	0	5

[a]Determined using antibodies CK2 and P3.58BA-14 in double-labelling procedure (Pantel et al 1991).
[b]$P<0.04$ as compared to adenomatous histology (χ^2 test).
[c]$P<0.05$ as compared to ICAM-1-negative tumours (χ^2 test).
[d]Mean time of observation: 13 months.

thickness of the primary tumour (the Breslow index) has remained such a good predictor of the eventual development of metastatic disease (Koh 1991).

Because of this strong correlation, the identification of phenotypic changes in the primary tumours which are associated with the onset of vertical growth might lead to the identification of markers of early metastatic cells. Towards this end, we have therefore isolated monoclonal antibodies which react with advanced primary tumours and metastases but which do not detect antigens in the early thin (<0.75 mm) tumours, most of which are in the radial growth phase. Using this approach, we have obtained several antibodies. When examined for reactivity with a large series of benign and malignant lesions, these antibodies react only rarely with melanocytes present in melanocytic or dysplastic nevi but are frequently positive on melanocytes in metastatic lesions. Among primary tumours, reactivity was found to increase in frequency and strength with increasing vertical thickness. The antigens defined by such antibodies therefore demonstrate an expression pattern which correlates with the development of metastatic potential. Such molecules might in fact contribute to metastasis and it is therefore important to try to obtain information on their functions. One approach to this is to isolate the cDNAs encoding these molecules since, from a comparison to molecules with proven functional domains, it is frequently possible to predict the function of a newly identified molecule. So far, we have cloned cDNAs for two of these molecules. Interestingly, both appear to be cell adhesion molecules and furthermore may mediate interactions between the tumour cells and leukocytes. cDNA cloning of the P3.58 antigen revealed that it is identical to ICAM-1 (Johnson et al 1988, 1989), an inducible cell adhesion molecule which has three known ligands ($\alpha_L\beta_2$, $\alpha_M\beta_2$ and CD43) all of which are expressed by activated leukocytes. The role of ICAM-1 is thought to be to strengthen interactions between effector cells and their targets (Springer 1990). The expression of this molecule by a tumour cell might therefore be expected to enhance immune recognition of the tumour and thus be associated with a good clinical prognosis. Although the expression of this molecule on some carcinomas is associated with parameters of good prognosis (Tomita et al 1990, and see above), the reactivity of the P3.58 monoclonal antibody with specimens of cutaneous melanoma suggested that an increase in antigen expression correlates with tumour progression. Independent investigations using another monoclonal antibody showed that the expression of ICAM-1 by Stage I melanomas is indeed associated with decreased disease-free interval and decreased survival time (Natali et al 1990). More recent studies, carried out with a series of monoclonal antibodies directed against different epitopes on the ICAM-1 molecule, indicate that not only is ICAM-1 expressed by benign as well as malignant melanocytic lesions but also that there were no differences in intensity of staining. This suggests that the antibodies which show a difference in reactivity between benign and malignant lesions reflect differences in the availability of particular epitopes between these lesions. ICAM-1 is a highly glycosylated molecule which shows great variability in degree of *N*-linked glycosylation between different cell types (Holzmann et al 1988). While the nature of these differences is not yet known, they may represent differences in post-translational modifications such as glycosylation, differences which could modify the adhesive activity of ICAM-1 (Diamond et al 1991).

MUC18 is a single-chain cell surface glycoprotein with an apparent molecular weight of 113 kDa (Lehmann et al 1987). Analysis of the MUC18 cDNA revealed that this molecule has all the characteristics of an integral membrane protein and that it is a member of the immunoglobulin supergene family (Lehmann et al 1989). Although MUC18 is a novel member of this family, it is related to several other molecules, all of which have been shown to mediate intercellular adhesion. The highest sequence similarity is to BEN/SC1/DM-GRASP (Pourquie et al 1992), a developmentally regulated cell adhesion molecule that plays a role in the development of the nervous and haemopoietic systems. Although BEN has thus far been identified only in the chicken, the extent of sequence similarity would not indicate that it is a homologue of MUC18. MUC18 is also related to DCC, a putative tumour suppressor gene in colorectal carcinoma

(Fearon et al 1990), the carcinoembryonic antigen family (Benchimol et al 1989) and to a group of molecules (NCAM, MAG, L1) which mediate cell–cell adhesion during the development of the nervous system (Edelman & Crossin 1991). Because of its sequence similarity to these molecules, each of which has been shown to mediate intercellular adhesion, it seems likely that MUC18 also is a cell adhesion molecule. Transfection of MUC18 cDNA into MUC18-negative cells led to strong surface expression of the glycoprotein but did not lead to an increase in aggregation between the cells (Johnson et al 1993). If MUC18 functions as a cell adhesion molecule it is likely therefore to be a heterophilic cell adhesion molecule interacting with a non-MUC18 ligand. While MUC18 is strongly expressed on the majority of melanomas, expression of the molecule in normal tissues is primarily on the vasculature where both smooth muscle cells and occasionally endothelial cells have been shown to be positive (Sers et al 1994). Analysis of the promoter region of the human gene (Sers et al 1993) revealed potential binding sites for a variety of transcriptional regulators and studies on glioma cell lines confirmed that, indeed, the expression of MUC18 can be modulated by agents which raise cAMP levels. Taken together, these observations indicate that MUC18 is an inducible endothelial cell surface molecule which may function in cell adhesion and suggest that its natural ligand may be found on a particular leukocyte subpopulation. Thus both molecules, which have been identified by monoclonal antibodies reacting more strongly with malignant than benign melanocytic tumours, are directed against cell surface molecules whose normal function appears to be to mediate leukocyte adhesion. The adhesion or aggregation of tumour cells with leukocytes has been shown to enhance tumour growth and metastasis in a number of animal models (Starkey et al 1984), a finding which is thought to reflect not only the production of growth factors and proteolytic enzymes by the leukocytes but also their role in mediating adhesion to the endothelium and migration into the tissues.

Changes in cell adhesion have long been known to be involved in the development of metastasis (Johnson 1991). The down-regulation of normally expressed cell adhesion molecules is an important early step in freeing cells from a variety of contact-mediated regulatory controls and allowing dissemination of cells from the primary tumour. Such a role has been postulated for DCC in colorectal tumours. In addition to the loss of normal cell–cell interactions, the metastasizing cell must interact with a variety of new cellular partners as it travels throughout the body and establishes secondary sites of growth (Hart et al 1989). Such new cell interactions, particularly in the vascular and lymphatic systems, may well be determined and guided by the expression of new or altered cell adhesion molecules. Why alterations in cell adhesion molecule expression occur during tumorigenesis remains one of the most important unanswered questions. While this may, as in the case of DCC, sometimes be due to genetic changes, there is no evidence that this is a common mechanism. Cell adhesion molecules are developmentally and environmentally regulated molecules

(Edelman & Crossin 1991) and it seems likely that their expression pattern during tumour progression reflects a disturbance at the level of the molecular elements normally responsible for controlling their expression. For most cell adhesion molecules, these elements remain undefined. The identification of these mechanisms may shed light not only on the normal expression of these molecules, but also on the changes which occur during the development of metastatic potential in human tumours.

Acknowledgement

This work was supported by a grant from the Dr Mildred Scheel Stiftung, Deutsche Krebshilfe, Bonn, Germany.

References

Benchimol S, Fuks A, Jothy S, Beauchemin N, Shirota K, Stanners CD 1989 Carcinoembryonic antigen, a human tumor marker, functions as an intercellular adhesion molecule. Cell 57:327–334

Brugger W, Bross KJ, Glatt M et al 1994 Mobilization of tumor cells and haematopoietic progenitor cells into peripheral blood of patients with solid tumors. Blood 83:636

Clark WH, Elder DE, Guerry D et al 1989 Model predicting survival in stage I melanoma based on tumour progression. J Natl Cancer Inst 81:1893–1904

Cote RJ, Rosen PP, Lesser ML et al 1991 Prediction of early relapse in patients with operable breast cancer by detection of occult bone marrow micrometastases. J Clin Oncol 9:1749

Delsol G, Gatter KC, Stein H et al 1984 Human lymphoid cells express epithelial membrane antigen. Lancet 2:1124–1128

Diamond MS, Staunton DE, Marlin SD, Springer TA 1991 Binding of the integrin Mac-1 (CD11b/18) to the third immunoglobulin-like domain of ICAM-1 (CD54) and its regulation by glycosylation. Cell 65:961–971

Diel IJ, Kaufmann M, Goemer R et al 1992 Detection of tumor cells in bone marrow of patients with primary breast cancer: a prognostic factor for distant metastasis. J Clin Oncol 10:1534

Edelman GM, Crossin KL 1991 Cell adhesion molecules: implications for a molecular histology. Ann Rev Biochem 60:155–190

Fearon ER, Cho KR, Nigro JM et al 1990 Identification of a chromosome 18q gene that is altered in colorectal cancers. Science 247:49–56

Fidler IJ, Radinski R 1990 Genetic control of cancer metastasis. J Natl Cancer Inst 82:166–168

Göttlinger HG, Funke I, Johnson J, Gokel JM, Riethmüller G 1986 The epithelial cell surface antigen 17-1A, a target for antibody-mediated tumor therapy: its biochemical nature, tissue distribution and recognition by different monoclonal antibodies. Int J Cancer 38:47–53

Hämmerling G, Maschek V, Sturmhöfel K, Momburg F 1989 Regulation and functional role of MHC expression on tumors. In: Melchers F et al (eds) Progress in immunology. Berlin: Springer p 1071–1078

Hart IR, Goode NT, Wilson RE 1989 Molecular aspects of the metastatic cascade. Biochim Biophys Acta 989:65–84

Herlyn DM, Steplewski Z, Herlyn MF, Koprowski H 1980 Inhibition of growth of colorectal carcinoma in nude mice by monoclonal antibody. Cancer Res 40:717–721

Holzmann B, Lehmann JM, Ziegler-Heitbrock HWL, Funke I, Riethmüller G, Johnson JP 1988 Glycoprotein P3.58, associated with tumor progression in malignant melanoma, is a novel leukocyte activation antigen. Int J Cancer 41:542–547

Jahn L, Fourquet B, Rohe K et al 1987 Cytokeratins in certain endothelial and smooth muscle cells of two taxonomically distant vertebrate species, *Xenopus laevis* and man. Differentiation 36:234–254

Järvinen M, Andersson LC, Virtanen I 1990 K562 erythro-leukemic cells express cytokeratins 8, 18 and 19 epithelial membrane antigen that disappear after induced differentiation. J Cell Physiol 143:310–320

Johnson JP 1991 Cell adhesion molecules of the immunoglobulin supergene family and their role in malignant transformation and progression to metastatic disease. Cancer Metas Rev 10:11–22

Johnson JP, Stade BG, Hupke U, Holzmann B, Riethmüller G 1988 The melanoma progression associated antigen P3.58 is identical to the intercellular adhesion molecule ICAM-1. Immunobiol 178:275–284

Johnson JP, Stade BG, Holzmann B, Schwäble W, Riethmüller G 1989 *De novo* expression of cell adhesion molecule ICAM-1 in melanoma and increased risk of metastasis. Proc Natl Acad Sci USA 86:641–644

Johnson JP, Rothbächer U, Sers C 1993 The progression associated antigen MUC18: a unique member of the immunoglobulin supergene family. Melanoma Res 3:337–340

Koh HK 1991 Cutaneous melanoma. New Engl J Med 325:171–182

Krook JE, Moertel CG, Gunderson LL et al 1991 Effective surgical adjuvant therapy for high-risk rectal carcinoma. N Engl J Med 324:709–715

Lehmann JM, Holzmann B, Breitbart EW, Schmiegelow P, Riethmüller G, Johnson JP 1987 Discrimination between benign and malignant cells of the melanocytic lineage by two novel antigens, a glyprotein with a molecular weight of 113 000 and a protein with a molecular weight of 76 000. Cancer Res 47:841–847

Lehmann JM, Riethmüller G, Johnson JP 1989 MUC18, a marker of tumor progression in human melanoma shows sequence similarity to the neural cell adhesion molecules of the immunoglobulin superfamily. Proc Natl Acad Sci USA 86:9891–9895

Lindemann F, Schlimok G, Dirschedl P, Witte J, Riethmüller G 1992 Prognostic significance of micrometastatic tumour cells in bone marrow of colorectal cancer patients. Lancet 340:685–689

Liotta LA, Steeg PS, Stetler-Stevenson WG 1991 Cancer metastasis and angiogenesis: an imbalance of positive and negative segulation. Cell 64:327–336

Litvinov SV, Velders MP, Bakker HAM, Fleuren GJ, Warnaar SO 1994 Ep-CAM: a human epithelial antigen is a homophilic cell–cell adhesion molecule. J Cell Biol 125:437–446

Moertel CG, Fleming TR, MacDonald JS et al 1991 Levamisole and fluoroacil for adjuvant therapy of resected colon carcinoma. N Engl J Med 322:352–358

Natali P, Nicotra MR, Cavaliere R et al 1990 Differential expression of intercellular adhesion molecule 1 in primary and metastatic melanoma lesions. Cancer Res 50:1271–1278

Pantel K, Schlimok G, Kutter D et al 1991 Frequent down-regulation of major histocompatibility class I antigen expression on individual micrometastatic carcinoma cells. Cancer Res 51:4712–4715

Pantel K, Angstwurm M, Kutter D et al 1992 Down-regulation and neo-expression of cell surface antigens on individual micrometastatic carcinoma cells. In: Rabes H, Peters PE, Munk K (eds). Metastasis: basic research and its clinical applications. Karger Verlag, Basel, Contrib Oncol Vol 44:283–293

Pantel K, Koprowski H, Riethmüller G 1993a Conference on cancer micrometastasis: biology, methodology and clinical significance. Int J Oncol 3:1019–1022

Pantel K, Schlimok G, Braun S et al 1993b Differential expression of proliferation-associated molecules in individual micrometastatic carcinoma cells. J Natl Cancer Inst 85:1419–1424

Pantel K, Izbicki JR, Angstwurm M et al 1993c Immunocytochemical detection of bone marrow micrometastasis in operable non-small cell lung cancer. Cancer Res 53:1027–1031

Pantel K, Schmaus W, Schlimok G, Riethmüller G 1993d p53 immunostaining of megakaryocytes is preferentially observed in patients with epithelial malignancies. Blood 82:610a (abstr)

Pantel K, Schlimok G, Angstwurm N et al 1994a Methodological analysis of immunocytochemical screening for disseminated epithelial tumor cells in bone marrow. J Hematother 3:165–173

Pantel K, Oberneder R, Klein C, Schlimok G, Riethmüller G 1994b In vitro expansion of micrometastatic carcinoma cells. Proc Am Assoc Cancer Res 35:60 (abstr)

Passlick B, Izbicki JR, Simmel et al 1994 Expression of major histocompatibility class I and class II antigens and intercellular adhesion molecule-1 on operable non-small cell lung carcinomas. Eur J Cancer 30:376

Peifer M 1993 Cancer, catenins, and cuticle pattern: a complex connection. Science 262:667

Pourquie O, Corbel C, La Caer JP, Rossier J, LeDouarin NM 1992 BEN, a surface molecule of the immunoglobulin superfamily expressed in a variety of developing systems. Proc Natl Acad Sci USA 89:5261–5265

Price JE, Aukerman SL, Ananthaswamy N et al 1989 Metastatic potential of cloned murine melanoma cells transfected with activated c-Ha-ras. Cancer Res 49: 4274–4281

Riesenberg R, Oberneder R, Kriegmair M et al 1993 Immunocytochemical double staining of cytokeratin and prostate specific antigen in individual prostatic tumour cells. Histochem 99:61–66

Riethmüller G, Johnson JP 1992 Monoclonal antibodies in the detection and therapy of micrometastatic epithelial cancers. Curr Opin Immunol 4:647–655

Riethmüller G, Schneider-Gädicke E, Johnson JP 1993 Monoclonal antibodies in cancer therapy. Curr Opin Immun 5:732–739

Riethmüller G, Schneider-Gädicke E, Schlimok G et al 1994 Randomised trial of monoclonal antibody for adjuvant therapy of resected Dukes' C colorectal carcinoma. Lancet 343:1177–1183

Schlimok G, Funke I, Holzmann B et al 1987 Micrometastatic cancer cells in bone marrow: in vitro detection with anti-cytokeratin and in vivo labelling with anti-17-1A monoclonal antibodies. Proc Natl Acad Sci USA 84:8672–8676

Schlimok G, Riethmüller G 1990 Detection, characterization and tumorigenicity of disseminated tumor cells in human bone marrow. Semin Cancer Biol 1:207–215

Schlimok G, Funke I, Pantel K et al 1991 Micrometastatic tumour cells in bone marrow of patients with gastric cancer: Methodological aspects of detection and prognostic significance. Eur J Cancer 27:1461–1465

Schlimok G, Lindemann F, Holzmann K, Witte J, Renner D, Riethmüller G 1992 Prognostic significance of disseminated tumor cells detected in bone marrow of patients with breast and colorectal cancer: a multivariate analysis. Proceedings ASCO 11:102 (abstr)

Schwarz M, Owasibe K, Kartenbeck J, Franke W 1990 Desmosomes and hemidesmosomes: constitutive molecular components. Ann Rev Cell Biol 6:461

Sers C, Kirsch K, Rothbächer U, Riethmüller G, Johnson JP 1993 Genomic organization of the melanoma-associated glycoprotein MUC18: implications for the evolution of the immunoglobulin domains. Proc Natl Acad Sci USA 90:8514–8518
Sers C, Riethmüller G, Johnson JP 1994 MUC18, a melanoma-progression associated molecule—a potential role in tumor vascularization and hematogenous spread? Cancer Res 54:5689–5694
Springer TA 1990 Adhesion receptors of the immune system. Nature 346:425–434
Starkey JR, Liggitt HD, Jones W, Hosick HL 1984 Influence of migratory blood cells on the attachment of tumor cells to vascular endothelium. Int J Cancer 34:535–543
Tanaka K, Yoshioka T, Bieberich C, Jay G 1988 Role of the major histocompatibility complex class I antigens in tumor growth and metastasis. Ann Rev Immunol 6:359–380
Tomita Y, Nishiyama T, Watanabe H, Fujiwara M, Sato S 1990 Expression of intercellular adhesion molecule-1 (ICAM-1) on renal-cell cancer: possible significance in host immune responses. Int J Cancer 46:1001–1006

DISCUSSION

Herrlich: Is it possible to do an adoptive bone marrow transplant with micrometastatic cells and prove that they are still able to form metastases?

Riethmüller: We have often tried this. So far we've been unable to do this in nude mice, but it has worked with oncogene-transformed cells that were expanded in bone marrow cultures using specific growth factors and extracellular matrix proteins.

Herrlich: Could you follow micrometastases of tumours using an isogenic animal system?

Riethmüller: There is no good model for early dissemination of epithelial tissue cells. The multiple intestinal neoplasia (MIN) mouse model has an *APC* deletion; this is the next model we would like to examine, to see whether micrometastatic cells can be detected in early stages of growing tumours.

Birchmeier: If the *17-1A* gene product is a homophilic cell adhesion molecule, and if you want to use the antibody for therapy, you have to assume that expression of this antigen is turned on during the progression of the tumour. Where would the homophilic adhesion then work? Is it adhesion in the micrometastases?

Riethmüller: We have tried very hard to show the putative adhesion function as described by Litvinov et al (1994). We also use transfectants. I think it may be a very borderline or weak homophilic interaction. Litvinov et al (1994) were able to show this in L cells, but we were unable to.

Pober: I'm curious as to your comment that the MUC18 antigen was also expressed by vascular smooth muscle cells. There are results that suggest that at least a subset of vascular smooth muscle cells, particularly in the aorta, derives embryologically from the neural crest (Beall & Rosenquist 1990, Hood & Rosenquist 1992). Is MUC18 expressed on smooth muscle cells throughout the body, or just in certain large arteries?

Riethmüller: I'm very interested to hear that. However, MUC18 is not only detectable on vascular smooth muscle, but it is also found on a variety of smooth muscle cells, including those of the gastrointestinal tract.

Hynes: You presented data about the overall homology of MUC18 with a number of immunoglobulin superfamily molecules. If you look specifically at its first domain, is it homologous with the active first domains of ICAM and VCAM? Does it have two disulphide bonds?

Riethmüller: Yes, there are two disulphide bonds in the first domain but this domain does not share high homology with the active first domains of ICAM and VCAM-1.

Hynes: I have a general point which seems to be suggested very clearly by your data. If these cells are already disseminated at an early stage, it raises serious questions about one's ability to block metastasis by blocking adhesion molecules.

Riethmüller: Yes, you are right: I think we will often be too late. In Dukes stage C colorectal carcinoma, for instance, about 40% of patients have tumour cells in bone marrow. The early-disseminated cells seem to have a different phenotype from those which later might have already metastasized from metastases. So the incidence of disseminated cells is low in early stages; I think these are cells mostly in the G0 phase of the cell cycle. I don't know why they emigrate from the primary tumour so soon. It may well be that they're not actually malignant: so far we haven't found them to be tumorigenic in nude mice.

Pober: Bernard Fisher made the point several years ago that in breast carcinomas there were often tumour cells circulating. The bad part about lymph node metastasis was not so much that the tumour cells were escaping from the primary breast carcinoma, because they did do very early, but that they had acquired the ability to grow in an ectopic location (reviewed in Fisher 1980).

Riethmüller: As to the microenvironment of metastases, we are now faced with the difficult task of making biopsy sections and analysing the vicinity of these cells to see what might be the signals that attract metastatic cells, and allow them to adhere and grow there.

Pober: But my point is that the outlook may not be as bad as you are portraying, because the cells that you are worried about may not yet have acquired the ability to grow elsewhere. Those that are in the bone marrow might not be the progenitors of the metastases elsewhere.

Hynes: The issue of whether you get secondary metastases from primary metastases is also involved here.

Riethmüller: Remember that many of the patients who have tumours arising from simple epithelia suffer from distant occult metastasis at the time of diagnosis of their primary tumour. Early dissemination of tumour cells is the major cause of death from cancer in Western nations.

Hynes: Presumably, blocking homing of those tumour cells will do you no good: it's all over by then.

Riethmüller: Nevertheless, we think that micrometastases are excellent targets for cytotoxic antibodies. For this purpose, we carried out a prospective randomized trial in which 200 patients were divided into two groups: in one group patients were treated with an antibody against the 17-1A antigen and, in the other, they were observed only (Riethmüller et al 1994). We followed up the patients for six years. To our astonishment, we found a clear difference. What was even more interesting was that the therapeutic effect was most pronounced on distant metastases. The rate of local relapses was not affected by the antibody treatment. We could reduce the mortality by 30% using this antibody with a rather low affinity. The problem with passive antibody treatment is accessibility of tumour cells. If you have a razor-sharp carving knife in your hand and you apply it to marble, it's no good, but apply it to linden wood and it's marvellous. Thus the question is, can we judge antibody therapy if we have applied it only to large terminal-stage tumours where the cells are hardly accessible to the injected antibody?

Hynes: That's very interesting. You don't know how the antibody is working: you don't know for sure that you're blocking homing. You may well be doing something else altogether, but it doesn't matter for the patients!

Riethmüller: The present phase of my clinical work is aimed at establishing a surrogate marker. I think the reason why the progress in therapeutic oncology is so scandalously slow is because oncologists lack a surrogate marker for adjuvant therapy. We have been concentrating on the rare patients who have a hundred or a thousand tumour cells per million bone marrow cells. We injected cytotoxic antibodies and, after the last infusion, we aspirated the marrow and counted the tumour cells again. And, indeed, one can show a definite decrease in those patients who express the target antigen. There was no decrease of cells in either the patients who got a placebo or in those patients who received the antibody but did not express the antigen on their tumour cells. Thus I'm very confident that one day we will establish a good surrogate model to study the question of the reaction mechanism directly in the patient.

Pober: If you remove the primary and look at the bone marrow before and after its removal, is there a drop in the number of tumour cells found in bone marrow? Could this population represent tumour cells that have recently arrived and are not capable of surviving?

Riethmüller: We have done a trial and studied the disseminated cells before and after surgery. There was no difference. As to longitudinal follow up in individual patients, there seems to be a rather tumour specific behaviour. In mammary as well as colorectal carcinoma, we have seen patients who have been repeatedly analysed and, who, after removal of the primary tumour, stay positive. The pathologists always argue that these isolated cells are not malignant cells. Yet the evidence that has piled up is that these early disseminated cells are in fact a good indicator of the disseminative capacity of the individual tumour.

References

Beall AC, Rosenquist TH 1990 Smooth muscle cells of neural crest origin form the corticopulmonary septum in the avian embryo. Anat Rec 226:360–366

Fisher B 1980 Laboratory and clinical research in breast cancer—a personal adventure. Cancer Res 40:3863–3874

Hood LC, Rosenquist TH 1992 Coronary artery development in the chick: origin and development of smooth muscle cells and the effect of neural crest ablation. Anat Rec 234:391–400

Litvinov SV, Velders MP, Bakker HAM, Fleuren GJ, Warnaar SO 1994 Ep-CAM: a human epithelial antigen is a homophilic cell–cell adhesion molecule. J Cell Biol 125:437–446

Riethmüller G, Schneider-Gädicke E, Schilmok G et al 1994 Randomised trial of monoclonal antibody for adjuvant therapy of resected Dukes' C colorectal carcinoma. Lancet 343:1177–1183

General discussion II

Haskard: We have heard a lot at this meeting about the importance for future therapy of obtaining a true quantitative understanding of the expression of adhesion molecules in disease. We have also heard about the present reliance on immunohistological techniques for defining adhesion molecule expression *in situ* and the well recognized problems of using those techniques quantitatively. Jonathan Barker has also brought up the issue of the difficulty in obtaining tissue from internal organs of living patients.

With these considerations in mind, we have been interested in the possibility of developing techniques for imaging adhesion molecule expression non-invasively. In the first instance we have directed our attention to E-selectin, which we chose to target because of its specific expression on activated endothelial cells. The rationale behind this work was very simple: if a circulating leukocyte could identify ligands on endothelial cells, so too could a radiolabelled monoclonal antibody.

We had developed an antibody (1.2B6) that reacts with pig E-selectin. Before we started imaging, we validated that it was possible to target activated endothelium *in vivo* with the radiolabelled antibody and that any tissue localization that we achieved was not simply due to permeability changes in the endothelium allowing non-specific exudation of immunoglobulin. This validation was done using a pig model in which inflammatory lesions of various durations were set up by injecting the skin with IL-1 (Keelan et al 1994a). The lesions were initiated at 24 h, 4 h, 2 h and 45 min before the end of the experiment and 10 min before the end of the experiment we injected 99m-Tc-labelled 1.2B6 monoclonal antibody and ^{111}In-labelled control antibody MOPC21. At the end of the experiment the skin spots were excised and radioactivity was counted.

In unstimulated skin there was an increased uptake of 1.2B6 compared with MOPC21, consistent with the constitutive expression of E-selectin in pig skin that can be detected immunohistologically. This is probably a species difference, as endothelial cells in normal human skin express minimal E-selectin (Cotran et al 1986, Norris et al 1991). The constitutive expression of E-selectin in pig skin is in contrast to other pig tissues which do not express the molecule unless stimulated.

We found that there was clearly increased uptake of anti-E-selectin following injection of IL-1, with maximum localization in lesions of 2 h and 4 h duration. We could also derive dose–response data using this approach with less uptake of antibody in lesions stimulated with smaller amounts of IL-1. There was very little uptake of MOPC21 in any of the skin spots, showing that the localization of anti-E-selectin was specific.

Having shown that we could successfully target E-selectin with the radiolabelled antibody, we chose to test its imaging capacity in a model of monoarthritis caused by injection into the knee of phytohaemagluttin (PHA) (Keelan et al 1994b). Following initiation of the arthritis, ^{111}In-labelled 1.2B6 or ^{111}In-labelled control MOPC21 was injected intravenously and gamma camera images collected over 24 h. We found that there was specific imaging with the anti-E-selectin monoclonal antibody of the inflamed knee and also of the draining regional lymph node. Furthermore, the outline of the torso was imaged, consistent with uptake by the constitutively expressed E-selectin in the skin.

Having established that it is possible to image activated endothelium in this way, we plan to progress in two directions. First, we hope to start using this technique in patients. Secondly, we should like to scale down to smaller laboratory animals in order to study more complex animal models of disease.

Finally, I should like to acknowledge the major contributions to this work of Drs Ted Keelan, Peter Chapman and Andrew Harrison in my laboratory at the RPMS, of Drs Mike Peters and Francois Jamar in the Department of Nuclear Medicine at the RPMS and of Dr Richard Binns and Steve Licence at the AFRC Babraham Institute.

Rothlein: Do pigs have circulating E-selectin?

Haskard: Yes, we have an assay that can detect that. We have some information that it goes up in inflammation. The question then arises as to whether circulating E-selectin would affect the capacity to image by non-specific blockade of the antibody. I suspect, however, that it will be to our advantage in that the circulating antigen may lead to very rapid clearance of the background radioactivity and one will be left with a more readily detected image.

Winn: Do you plan to use the control antibody at the same time in humans?

Haskard: In humans it's more likely that we'll use differentially radiolabelled neutrophils at the same time. Radiolabelled neutrophils usually tell you that there's infection. They are quite specific for infection but rather insensitive. We have data from our pig model that the antibody is probably going to be more sensitive at picking up inflammatory lesions than radiolabelled neutrophils.

Herrlich: What is the message that is transferred to the local lymph node?

Haskard: I don't know the answer to that. In the model that we've used we've injected PHA; it could be PHA itself or it could be a cytokine that PHA is inducing. But we also have another model in which we inject urate crystals as a model of gout, and we also see lymph node activation in that model.

Elices: I wanted to ask you about the technological background of this imaging technique. What are your goals for the future in terms of persuing this as a potential diagnostic tool? Do you see it as using radioactivity for imaging, or do you envisage labelling with fluorescence?

Haskard: Practically, it would have to be radioactivity. The amount of radioactivity involved is not great and the aim would be to use it for a variety of clinical situations, not only for initial diagnostic purposes but possibly also for following patients. There are a number of clinical situations where it would be nice to have a handle on what's going on in the inflammatory response—these include inflammatory bowel disease, cerebral lupus, psoriasis and rheumatoid arthritis.

Stanley: How does this compare with a gallium scan?

Haskard: A gallium scan is rather insensitive and takes two to three days. Gallium is thought to be taken up by macrophages in granulomas. One of the problems is that the background takes a long time to clear. If you want a quick clinical answer that's not too convenient. We can get an image in 3 h which potentially is an improvement.

Rothlein: If you're going to follow an inflammation as you treat it, have you tried re-treating the pig yet?

Haskard: We did try that in the very early days of the experiments. When we did a repeat experiment after 10 days we found that all the tracer went into the lungs. This was probably due to the presence of pig anti-mouse immunoglogulin and the clearance of immune complexes by the lungs.

Rothlein: So you still have to overcome that obstacle.

Haskard: Sure, we would need to do that if we were to use the techniquc serially. The aim would be to make less immunogenic forms of the antibody in the form of sFv fragments or 'humanized' material. I acknowledge that these will not necessarily be non-immunogenic but they might be an improvement.

Winn: With the 'humanized' version you might have a half-life of several days and you may not even have to treat twice.

Haskard: Clinically, a long half-life is not going to be to one's advantage because it's going to lead to a higher background.

References

Cotran RS, Gimbrone MA, Bevilacqua MP, Mendrick DL, Pober JS 1986 Induction and detection of a human endothelial activation antigen in vitro. J Exp Med 164:661–666

Keelan ETM, Licence ST, Peters AM, Binns RM, Haskard DO 1994a Characterization of E-selectin expression in vivo using a radiolabelled monoclonal antibody. Am J Physiol 266:H279–H290

Keelan ETM, Harrison AA, Chapman PT, Binns RM, Peers AM, Haskard DO 1994b Imaging vascular endothelial activation: an approach using radiolabelled monoclonal antibody against the endothelial cell adhesion molecule E-selectin. J Nucl Med 35:276–281

Norris P, Poston RN, Thomas DS, Thornhill M, Haw J, Haskard DO 1991 The expression of endothelial leukocyte adhesion molecule-1 (ELAM-1), intercellular adhesion molecule-1 (ICAM-1) and vascular cell adhesion molecule-1 (VCAM-1) in experimental cutaneous inflammation: a comparison of ultraviolet-B erythema and delayed hypersensitivity. J Invest Dermatol 96:763–770

Mechanisms of VCAM-1 and fibronectin binding to integrin $\alpha_4\beta_1$: implications for integrin function and rational drug design

Martin J. Humphries, Joe Sheridan, A. Paul Mould and Peter Newham

School of Biological Sciences, University of Manchester, 2.205 Stopford Building, Oxford Road, Manchester M13 9PT, UK

Abstract. Integrin $\alpha_4\beta_1$ can mediate both cell–cell and cell–extracellular matrix adhesion by binding to either fibronectin or vascular cell adhesion molecule 1 (VCAM-1). Both interactions are important for extravasation of leukocytes from the blood implying that rationally designed inhibitors of $\alpha_4\beta_1$ function may be useful for treating various inflammatory conditions. The mechanisms of ligand binding by $\alpha_4\beta_1$ are complicated by the fact that alternative splicing can generate different isoforms of the receptor-binding domains in both fibronectin and VCAM-1. Therefore, in addition to developing $\alpha_4\beta_1$ antagonists, we have also been interested in identifying isoform-specific functions. Recombinant ligand variants have been tested in adhesion and direct receptor-binding assays and each molecule was found to have a different inherent affinity for $\alpha_4\beta_1$ that endows them with different adhesive activities. This suggests that alternative splicing may regulate $\alpha_4\beta_1$-dependent motility *in vivo*. The initial strategy that we have adopted to develop $\alpha_4\beta_1$ inhibitors has been to identify key amino acid residues and peptide sequences participating in the receptor–ligand binding event and to use this information to generate synthetic mimetics. Three active sites have been identified in fibronectin by testing truncated proteins, expressing recombinant fragments and screening synthetic peptides. Two of these sites employ versions of a novel integrin-binding motif, LDVP/IDAP. A key active site in VCAM-1 has been identified by similar approaches as the related sequence IDSP. Since IDSP-like sequences are probably used by other integrin-binding immunoglobulins, derivatives of these peptides may turn out to be the forerunners of a new generation of therapeutic agents with multiple applications.

1995 Cell adhesion and human disease. Wiley, Chichester (Ciba Foundation Symposium 189) p 177–194

Adhesion is a fundamental prerequisite for many normal cell functions. This need is most commonly exemplified by the dependence of growth on anchorage, but adhesion also generates the traction required for cell movement and cell positioning, and directly influences intracellular signalling systems and gene expression. In addition to normal functions, adhesion also contributes to the pathogenesis of many of the most common human diseases. In these cases, if cell-adhesive interactions could be regulated, then disease progression might be arrested. Prime molecular targets for the generation of adhesion regulators are adhesion receptors, of which integrins are the principal class. Integrins are a family of α, β heterodimeric glycoproteins that currently contains 21 different members in vertebrates (Hynes 1992). Each dimer possesses a unique ligand specificity, suggesting that different receptors exhibit different functions *in vivo* and implying that specific integrin antagonists could have many applications (Humphries 1990).

The first step in the rational design of adhesion regulators requires information about the structure of adhesion molecules. Ideally, this would be a three-dimensional description of the receptor–ligand binding complex, but, alternatively, active site domains of either the receptor or ligand could be studied in isolation. In recent years, the rational design of integrin antagonists has been aided immensely by studies of the molecular basis of ligand binding. A key concept that has emerged from this work is that the majority of integrin ligands use very short protein sequences as receptor recognition motifs (Humphries 1990, Yamada 1991). Analogues of these sequences are essentially lead compounds for drug design (Humphries et al 1994). Initial studies, using fibronectin as a model, mapped adhesive domains using proteolytic fragments to the point at which it was possible to reproduce activity in the form of synthetic peptides. This important strategic transition permitted extremely rapid elucidation of a key minimal active sequence as the tripeptide RGD (Pierschbacher & Ruoslahti 1984). The discovery that peptides containing such a short sequence could possess adhesive activity has since been substantiated by extensive antibody and mutagenesis studies of fibronectin, and by the fact that other adhesion molecules, including fibrinogen, von Willebrand factor, vitronectin and osteopontin have subsequently been shown to employ the same RGD motif, albeit with different receptor specificity profiles (Humphries 1990).

RGD-dependent binding of fibrinogen to the platelet integrin $\alpha_{IIb}\beta_3$ (also known as GPIIb/IIIa), an interaction that plays a key role in thrombosis, is the first integrin–ligand interaction to be targeted by pharmaceutical companies for drug development. This work has now led to the development of RGD mimetics which exhibit potency, specificity and a capability for oral delivery (Humphries et al 1994). Current indications from clinical trials with these antagonists in unstable angina, coronary angioplasty and myocardial infarction are encouraging. In addition to being used in soluble form, the RGD motif has also been incorporated into formulations with hydrophobic peptide linkers or glycosaminoglycans to enhance cell adhesion. Initial clinical trial data from the treatment of ulcers and tissue grafts have shown positive effects on the acceleration of wound closure, improvement of cellular organization and reduction of scarring.

Subsequent to the discovery of the RGD recognition sequence, the molecular basis of other integrin–ligand interactions has been partially solved. In three cases activity can again be described by synthetic peptides: first, proteolytic dissection of fibrinogen has led to the identification of a peptide from the C-terminus of the γ-chain with a minimal sequence QAGDV which binds to $\alpha_{IIb}\beta_3$ (Kloczewiak et al 1984); second, the major integrin $\alpha_4\beta_1$-binding site in fibronectin has been localized to the tripeptide LDV (Komoriya et al 1991); and third, evidence has been presented to implicate the sequence DGEA as an active site in collagen type I for binding to the integrin $\alpha_2\beta_1$ (also known as GPIa/IIa) (Staatz et al 1991). There appears to be a functional analogy between some of these motifs since QAGDV and RGD peptides bind to the same or mutually exclusive binding sites on $\alpha_{IIb}\beta_3$ (Santoro & Lawing 1987) and LDV and RGD have the same relationship for binding $\alpha_4\beta_1$ (Mould et al 1991). This suggests that the concepts learned from studies of RGD may be applicable to other integrin-binding motifs.

The integrin $\alpha_4\beta_1$ and its ligands

In recent years, work in this laboratory has focused on the leukocyte integrin $\alpha_4\beta_1$ (also known as VLA-4) and its two ligands, the extracellular matrix glycoprotein fibronectin (Wayner et al 1989, Guan & Hynes 1990, Mould et al 1990) and the endothelial cell-surface protein VCAM-1 (Elices et al 1990). Since anti-$\alpha_4\beta_1$ monoclonal antibodies have been shown to inhibit trafficking of leukocytes in a number of acute and chronic inflammatory conditions *in vivo*, it appears that these interactions contribute to leukocyte extravasation and are therefore important therapeutic targets (Yednock et al 1992, Weg et al 1993). Consequently, we have been interested in determining the molecular basis of $\alpha_4\beta_1$–ligand binding with a view to using this information to develop specific antagonists.

In addition to the role of $\alpha_4\beta_1$ in disease processes, results from several studies have also identified $\alpha_4\beta_1$–ligand interactions as a useful model system for studying the regulation of integrin-mediated migration: first, $\alpha_4\beta_1$ is generally expressed by highly motile cell types such as leukocytes and neural crest derivatives, and is employed during migration of these cells *in vitro* (Hemler et al 1990); second, expression of $\alpha_4\beta_1$ by tumour cells correlates with invasiveness and metastatic potential (Albelda et al 1990); third, expression of a chimeric integrin containing the α_2 extracellular domain and the α_4 cytoplasmic domain enhances the ability of cells to migrate on collagen substrates compared to cells transfected with chimaeras containing other cytoplasmic domains (Chan et al 1992). It is therefore conceivable that an understanding of the role of $\alpha_4\beta_1$ in cell migration will yield insights that are applicable to other adhesion systems.

$\alpha_4\beta_1$ binding to fibronectin occurs in a complex manner via the HepII/IIICS region. HepII is the major proteoglycan-binding domain of fibronectin, while

IIICS is one of three sites subject to alternative splicing within the molecule. Three distinct $\alpha_4\beta_1$-binding sites have been identified within the HepII/IIICS region (Humphries et al 1986, 1987, Mould & Humphries 1991). Two sites (represented by peptides CS-1 and CS-5) are present in independently spliced segments of the IIICS (Humphries et al 1986, 1987), while the other site (represented by the peptide H-1) is found in the HepII region and is therefore expressed in all fibronectin isoforms (Mould & Humphries 1991). As discussed above, the CS-1 peptide contains the tripeptide LDV as its minimal active site (Komoriya et al 1991). H-1 contains a related motif, IDA, while CS-5 incorporates a variant of RGD, REDV (Humphries et al 1986, Mould & Humphries 1991). On the basis of previous studies of the relative activities of the three active site peptides, CS-1 has approximately 20-fold greater activity than CS-5 and the activity of H-1 is similar to that of CS-5 (Mould & Humphries 1991, Mould et al 1991). VCAM-1 is a member of the immunoglobulin superfamily, consisting of six or seven immunoglobulin repeats. The sites in VCAM-1 recognized by $\alpha_4\beta_1$ lie within immunoglobulin domains I and IV (Vonderheide & Springer 1992). In the six-domain form of VCAM-1, domain IV is removed by alternative splicing.

Functional aspects of $\alpha_4\beta_1$–ligand binding

The location of integrin binding sites in alternatively spliced segments of both $\alpha_4\beta_1$ ligands suggests that splicing may regulate function. To test this possibility, we adopted the approach of generating recombinant versions of the HepII/IIICS splice variants. Four variants have been studied which contain the four possible combinations of the three active sites—CS-1, CS-5 and H-1. The H-120 variant contains all three known sites, H-89 contains CS-1 and H-1, H-95 contains CS-5 and H-1, and H-0 contains H-1 alone (Fig. 1). When tested in cell attachment and spreading assays, the relative activities of the different variants mirrored the relative activities of CS-1, CS-5 and H-1 peptides—i.e. if present, the CS-1 sequence dominated the activity of the HepII/IIICS region (Mould et al 1994). The difference in activity between H-95 and H-0 implies that the CS-5 site is active in the H-95 variant.

Interestingly, inhibition experiments with anti-integrin antibodies suggested that whereas adhesion to VCAM-1, H-120 and H-89 was almost exclusively mediated by $\alpha_4\beta_1$, some of the cell attachment and migration mediated by H-95 and H-0 was not due to integrin recognition. Instead, attachment was partially inhibited by heparin, suggesting that proteoglycans were also mediating cell adhesion and that proteoglycans and $\alpha_4\beta_1$ may cooperate in promoting attachment to the HepII domain. In contrast, we observed that heparin did not significantly inhibit attachment to H-120 and H-89, suggesting that proteoglycan binding does not influence recognition of this sequence by $\alpha_4\beta_1$.

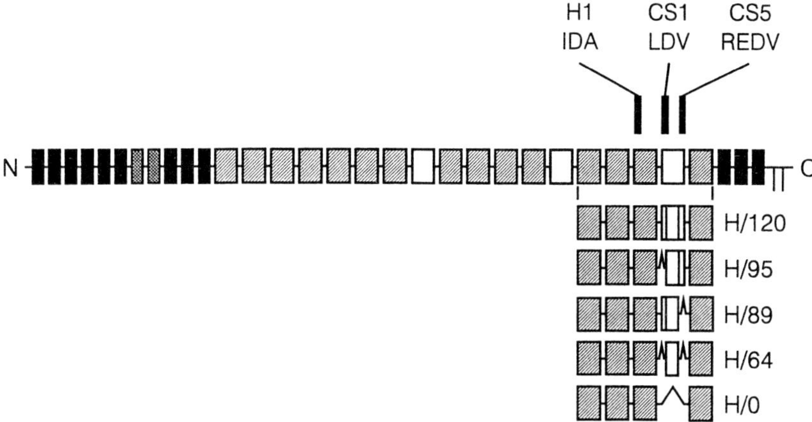

FIG. 1. Schematic representation of the four different recombinant variants of the HepII/IIICS region referred to in the text (H variants). The top part of the figure shows the modular structure of the fibronectin subunit: closed boxes, type I repeats; cross-hatched boxes, type II repeats; hatched boxes, type III repeats; open boxes, regions subject to alternative splicing. The methods used for polymerase chain reaction cloning and expression of H variants are given in Makarem et al (1994). The numbers assigned to H variants refer to the number of IIICS residues in the protein. The locations of the three recognition sequences for $\alpha_4\beta_1$, H-1, CS-1 and CS-5 are indicated.

Measurements of receptor binding affinities for the different recombinant proteins in a solid-phase binding assay in the presence of 1 mM $MnCl_2$ showed that VCAM-1 bound to $\alpha_4\beta_1$ with approximately fourfold higher affinity than H-120 and H-89 (which were both of similar affinity). H-120 exhibited approximately 10-fold higher binding affinity than H-95, which in turn exhibited approximately twofold higher affinity than H-0 (Mould et al 1994). These results match the observed differences between the ligands in adhesion assays. An important conclusion from these studies is that the effects of fibronectin and VCAM-1 on cell adhesion appear to depend only on the affinity of their interactions with $\alpha_4\beta_1$ and not on a ligand-specific sequence.

The significance of differences in affinity could be related to receptor reorganization with high-affinity ligand binding eliciting clustering of larger numbers of integrins compared to low-affinity ligand binding, and therefore to more stable interactions of integrins with the cytoskeleton. Alternatively, high-affinity ligand binding may transduce different signals to the cell interior compared to low-affinity binding, for example through effects on integrin conformation, clustering, or association of signalling molecules. *In vivo*, the high affinity binding of VCAM-1 to $\alpha_4\beta_1$ may favour the arrest of leukocytes on the surface of endothelial cells at sites of inflammation.

A novel finding from measurements of the kinetics of $\alpha_4\beta_1$–ligand binding was that changes in ligand affinity were due principally to changes in the rate of ligand–receptor association (k_1) rather than ligand–receptor dissociation (k_{-1}). The activation energies for these reactions are likely to derive mainly from conformational changes in both ligand and receptor; the identification of conformation-specific monoclonal antibodies has indicated that such changes do occur (Frelinger et al 1990, Zamarron et al 1991). The results from $\alpha_4\beta_1$–ligand binding assays suggest that receptor–ligand engagement requires changes in ligand conformation (these changes may only be small for sequences in VCAM-1, but very large for low-affinity sequences, such as in the HepII region of fibronectin). In contrast, receptor–ligand dissociation appears to require changes in receptor conformation.

Recombinant fibronectin and VCAM-1 ligands were also examined for their ability to promote melanoma cell migration. Migration on H-120, H-89 and VCAM-1 showed a biphasic response as a function of the coating concentration of ligand, with recombinant soluble VCAM-1 being two- to threefold more potent in the first phase. Higher concentrations of all ligands supported reduced cell migration. Our interpretation is that a threshold level of cell adhesiveness is needed for cell migration to start, at intermediate adhesiveness cell motility increases to a maximum and at high adhesiveness migration declines as a result of the inability of the cells to break contact with the substrate. Because adhesiveness is likely to be directly proportional to the affinity of receptor–ligand binding, the results can again be explained by the differences in ligand affinities. For H-95, which promotes weak migration that is not reduced at high coating concentration, it appears that its adhesive strength is too weak for cell migration to reach a maximum and, for H-0, the threshold level at which migration can begin is not attained.

We previously hypothesized that adhesion to lower-affinity sites, such as CS-5, may allow more rapid cell migration, whereas adhesion to the high-affinity CS-1 site may restrict cell migration. The results presented here suggest the opposite and point to a key role for CS-1 in migration. However, the concentration of fibronectin in the extracellular matrix is likely to be of crucial importance *in vivo*. If high concentrations of fibronectin containing the CS-1 site are present, the attachment strength may exceed that required for optimal cell migration. Enforced inclusion of different spliced segments of the IIICS region through transgenic approaches may provide additional insights into the effects of the different $\alpha_4\beta_1$ recognition sites on migration *in vivo*.

VCAM-1 active sites

As described above, VCAM-1-mediated leukocyte extravasation is an important therapeutic target for the development of anti-inflammatory agents. Recently, we have shown that fibronectin and VCAM-1 act as competitive inhibitors of each other's binding to $\alpha_4\beta_1$ (Makarem et al 1994). This suggests that $\alpha_4\beta_1$ may interact with fibronectin and VCAM-1 by a similar mechanism. The tripeptide sequence LDV is the key active sequence within CS-1 (Komoriya et al 1991) and, on inspection, VCAM-1 domains I and IV contain a related sequence, I(39)DSP. A computer-generated model of VCAM-1 domain I, based on immunoglobulin domain crystal structures, places this sequence in the exposed C–D loop region that is potentially accessible to $\alpha_4\beta_1$ (Fig. 2). Molecular dynamics simulations predict that the IDSP loop is flexible and projected away from the side of the domain.

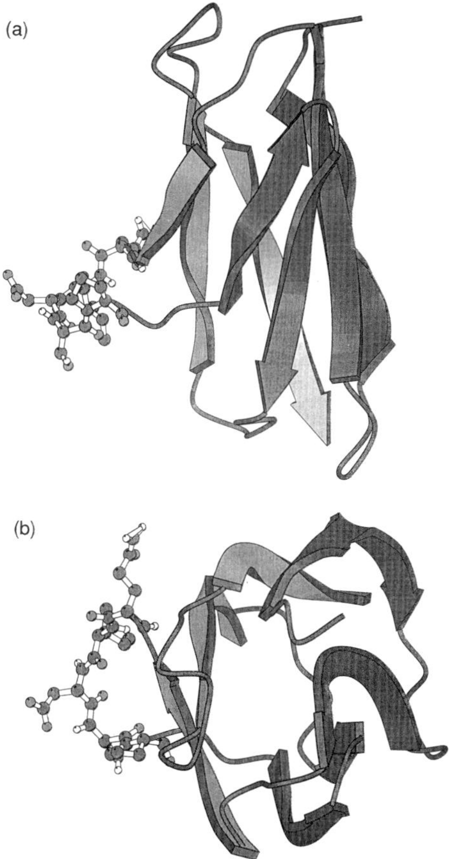

FIG. 2. Model of VCAM-1 domain I. The model shown is an averaged structure following molecular dynamics simulations and is based on the Ig crystal structure of 1REIa (for further details see Clements et al 1994). The diagram was produced with the aid of the display program MOLSCRIPT (Kraulis 1991). The β-strands that collectively form the ABDE sheet are shown as dark grey arrows and those that form the CFG sheet are shown as light grey arrows. The I(39)DSP sequence is shown in ball-and-stick form. In (a) the domain is viewed from the side; in (b) it is viewed from the top.

A number of mutations within the IDSP-containing loop and within other sequences predicted to be surface-accessible, were screened for their effects on adhesive activity. Using a simplified VCAM-1 construct lacking domains IV–VI, mutation of the I(39)DSP sequence essentially abrogated $\alpha_4\beta_1$–VCAM-1 binding (Clements et al 1994). The profile of monoclonal antibody binding to mutated VCAM-1 molecules was not affected by the mutation, indicating that the effects on adhesion were a direct consequence of perturbing integrin recognition. To examine the contribution of the IDSP sequence in domain IV of VCAM-1, we made use of the fact that $\alpha_4\beta_1$ adhesion to VCAM-1 exhibits temperature sensitivity; binding to domain I will take place at either 37 °C or 4 °C, while binding to domain IV only takes place at 37 °C (Needham et al 1994). Mutations in full length VCAM-1 within the IDSP region of domain I only partially reduced adhesion at 37 °C (Clements et al 1994). When adhesion was tested at 4 °C, almost all binding was abolished, indicating that the residual activity observed at 37 °C could be attributed to $\alpha_4\beta_1$ interaction with domain IV. Mutations of the I(327)DSP sequence of domain IV had no significant effect at either assay temperature, indicating that if I(327)DSP is active, $\alpha_4\beta_1$ preferentially binds domain I over domain IV. Simultaneous mutation of both IDSP sequences in domains I and IV produced a molecule that was unable to support adhesion at either temperature. Taken together, these data suggest that $\alpha_4\beta_1$ binds domain I and domain IV via a common mechanism mediated by the IDSP motif. However, if binding to one domain is compromised, binding can still take place via the other domain.

In agreement with the data from mutagenesis experiments, a peptide spanning the IDSP region (TQIDSPLN) was shown to inhibit spreading of melanoma cells on recombinant soluble VCAM-1 in a dose-dependent manner (Clements et al 1994). The peptide possessed similar activity to the fibronectin CS-1 peptide, while a control peptide with scrambled sequence (QNLSPITD) had no significant inhibitory effect. A peptide containing IDSP can therefore specifically and effectively perturb $\alpha_4\beta_1$–VCAM-1 interactions, thereby confirming the results obtained from VCAM-1 mutagenesis. Importantly, this result suggests that the mechanism of integrin binding by immunoglobulin ligands is related to that employed by most extracellular matrix ligands.

Although much structure–activity work remains to be performed, the IDSP motif is the third LDV-like sequence to be identified as a functional integrin-binding site (after the prototypic LDV(P) site in CS-1 and the IDA(P) sequence in the fibronectin HepII peptide H-1). A series of other reports suggest that there may be many more examples of the use of similar sequences and that this is in fact a second common integrin-binding motif, functionally related to RGD. One such sequence is the QAGDV peptide from the C-terminus of the γ-chain of fibrinogen (Kloczewiak et al 1984). This peptide bears some homology to LDV (Table 1) and, like LDV (Humphries et al 1987), is a competitive inhibitor of RGD function (Santoro & Lawing 1987). It is most striking, however, that

TABLE 1 Sequence homology in domain I C–D loop region of integrin-binding cell adhesion molecules and comparison with functionally active sequences in fibronectin and fibrinogen.

Protein	*Species*	*Sequence*
FN CS-1	Hu/Ra/Bo	PEI **LDVP** STV
FN CS-1	Ch	PDM **LDVP** SVD
FN CS-1	Xe	PEI **LDVP** TDE
FN H-1	Hu/Ra/Bo/Ch	STA **IDAP** SNL
FN H-1	Xe	TTA **VDSP** SNL
FG γ-chain	Hu	KQA **GDV**.
VCAM-1 DI	Hu/Ra	RTQ **IDSP** LNG
VCAM-1 DI	Mu	RTQ **IDSP** LNA
VCAM-1 DIV	Hu	RTQ **IDSP** LSG
VCAM-1 DIV	Mu/Ra	RTQ **TDSP** LNG
MAdCAM-1 DI	Hu	WRG **LDTS** LGS
ICAM-1 DI	Hu	LLG **IETP** LPK
ICAM-1 DI	Mu	SLG **LETQ** WLK
ICAM-1 DI	Ra	GLG **LETN** WMK
ICAM-2 DI	Hu	VGG **LETS** LNK
ICAM-2 DI	Mu	MGG **LETP** TNK
ICAM-3 DI	Hu	KIA **LETS** LSK

'LDVP' homology is in bold. FN, fibronectin; FG, fibrinogen; Hu, human; Ra, rat; Bo, bovine; Ch, chicken; Xe, *Xenopus*; Mu, mouse.

all other integrin-binding immunoglobulin ligands contain either an aspartate or a glutamate residue in their membrane-distal domains in a position analogous to the IDSP of VCAM-1 (Table 1). There is now considerable evidence that each of these residues contributes to function: mutation of (i) E34 within ICAM-1 domain I (Staunton et al 1990) and (ii) E37 within ICAM-3 (D. Simmons, personal communication) substantially inhibits $\alpha_L\beta_2$ (LFA-1) binding; (iii) mutation of D41 within MAdCAM-1 inhibits $\alpha_4\beta_7$ binding (M. Briskin, personal communication); and (iv) a peptide spanning the L(39) ETP site of of ICAM-2 is an effective inhibitor of $\alpha_L\beta_2$ binding and is capable of a direct interaction with $\alpha_L\beta_2$ (Li et al 1993). Thus, a consensus integrin-binding motif on immunoglobulin cell adhesion molecules becomes apparent: aliphatic–aspartate or glutamate–serine or threonine–proline or hydrophilic. The use of a glutamate rather than an aspartate residue by ICAMs contrasts with the absolute dependence of RGD sequences on aspartate, but might reflect a slightly deeper active site pocket in β_2 integrins. In the future, it will be of pressing importance to test the ability of LDV-like ('LDV') peptides to perturb each of these integrin–ligand interactions using sensitive assays for adhesion, since this would have implications for the design of novel classes of anti-adhesive agent.

One question prompted by the common use of RGD- and LDV-like sequences is the extent to which each integrin–ligand interaction is dependent on a short acidic peptide motif. In the case of fibrillar collagen molecules, which bind to $\alpha_1\beta_1$ and $\alpha_2\beta_1$, linear sequences may not be involved. Using the adhesion of human chondrosarcoma cells to type II collagen as a model system, we have found that native and heat-denatured collagen molecules support adhesion by different mechanisms. Recognition of native type II collagen by $\alpha_2\beta_1$ requires a triple helical conformation and is not inhibited by linear CNBr fragments or by the DGEA peptide described by Staatz et al (1991). Recognition of denatured collagen, on the other hand, is mediated by a fibronectin–$\alpha_5\beta_1$ bridge, not by $\alpha_2\beta_1$, and is inhibited by CNBr fragments. The inability to reproduce the integrin-binding activity of type II collagen in the form of a linear peptide implies that collagens employ a different type of active site in their receptors to those ligands that use RGD- or LDV-like sequences. This appears to be the case, since in recent work (D. S. Tuckwell and M. J. Humphries, unpublished results), we have shown that the recombinant I-domain of α_2 interacts specifically with fibrillar collagen in a cation-dependent manner.

Future prospects: integrin-binding peptide potency and specificity

The fact that both RGD and LDV are used as common recognition signals provides a partial explanation for the ligand-binding promiscuity of integrins. It also explains why an integrin ligand can competitively inhibit the binding of other ligands to the same integrin, e.g. fibrinogen, fibronectin and von Willebrand factor for $\alpha_{IIb}\beta_3$, and fibronectin and VCAM-1 for $\alpha_4\beta_1$. Some integrins, however, exhibit exquisite ligand-binding specificity, implying that either the integrins, the ligands, or both have structural features that generate this specificity. Little is known about integrin structure, but as far as the ligands are concerned, two possibilities are that the peptides themselves are held in different conformations in different ligands, or that other regions of the ligands contribute to receptor binding and endow specificity. Although there is some evidence supporting the latter view (Obara et al 1988, Bowditch et al 1994), it is now clear that simple alterations in the structure of synthetic RGD peptides can produce molecules that are selective inhibitors. In turn, this implies that normal flanking/spatially adjacent sequences in the ligand may serve to constrain RGD conformation. As yet, in the absence of crystal structure data, there are few indications as to how this is achieved. However, one reproducible finding from RGD structure–function studies has been the generation of greater potency in RGD peptides containing highly hydrophobic amino acids in the position following the aspartate (e.g. L, Y, F and W). This suggests that hydrophobic bonding may contribute to the interaction of RGD sequences with their integrin active sites. How this requirement relates to the hydrophobic residues in LDV motifs remains to be investigated. What is apparent so far from LDV structure–function studies is that the motif has greater tolerance of amino acid substitutions in the residues flanking the key aspartate. This offers hope that peptide engineering might produce selective inhibitors for cell adhesion molecule-binding integrins. As expected from work with RGD, linear LDV peptides are predicted to be extremely flexible (Fig. 3) and studies are now underway to introduce conformational restrictions into these peptides, initially by cyclization.

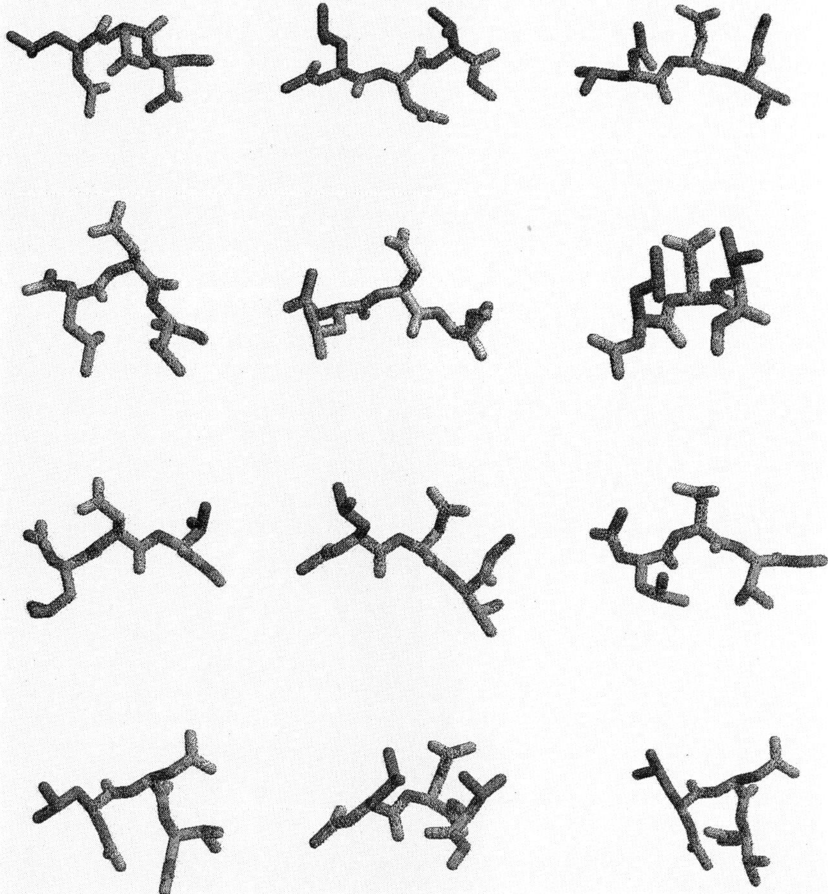

FIG. 3. Accessible low-energy conformers of Ac-LDV-N-methyl amide after molecular dynamics-simulated annealing at an exhaustively slow cooling rate (12.5 K/ps; J. Sheridan & M. J. Humphries, unpublished results). The high variability in low-energy structures demonstrates the flexibility of the linear peptide.

Tertiary structure determinations of RGD sites in fibronectin (Fig. 4), tenascin, kistrin, echistatin and recently in the foot-and-mouth disease VP1 coat protein have in all cases localized them to flexible surface loops (Adler et al 1991, Chen et al 1991, Leahy et al 1992, Main et al 1992, Logan et al 1993). Interestingly, although we cannot comment on flexibility, our prediction is that LDV active sites in integrin-binding immunoglobulins are also localized to loops. This location may be functionally important in endowing a relatively high on-rate for ligand binding or may reflect the fact that the majority of the loop is not anchored to the rest of the molecule, but it also suggests that there is likely to be a high degree of induced fit in the ligand subsequent to binding. Elucidation of the conformational changes that take place in both ligands and receptors during binding and dissociation, and how this varies between ligand isoforms, remains a fundamentally important question. In addition, further investigation of the molecular basis and functional consequences of ligand binding by the integrin $\alpha_4\beta_1$ promises to answer a number of other key questions.

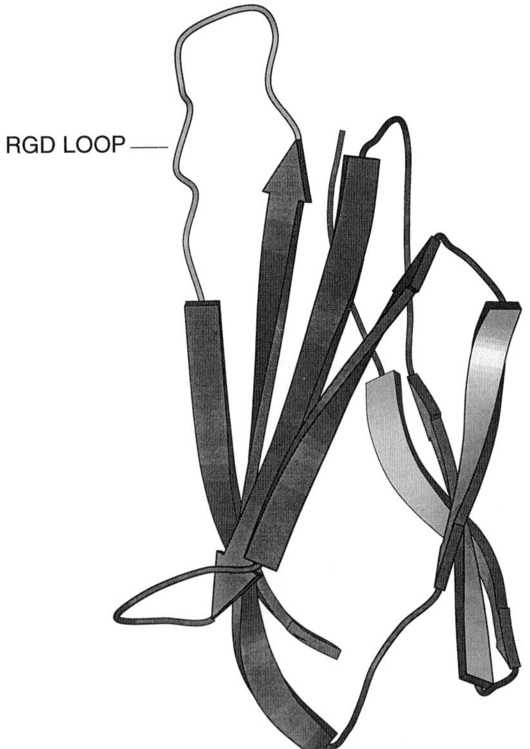

FIG. 4. Model of fibronectin type III repeat 10. The model is derived from atomic coordinates based on results from nuclear magnetic resonance studies of the recombinant molecule (Main et al 1992). The location of the RGD site is indicated.

For example, how large is the family of LDV-containing proteins? What are the essential structural features of an active 'LDV' sequence and can these be exploited to permit development of a novel family of specific integrin antagonists? What is the functional role of the short aspartate- or glutamate-containing peptides employed by integrin ligands and how is this related to the nature of integrin active sites?

References

Adler M, Lazarus RA, Dennis MS, Wagner G 1991 Solution structure of kistrin, a potent platelet aggregation inhibitor and GP IIb-IIIa antagonist. Science 253:445–448

Albelda SM, Mette SA, Elder DE et al 1990 Integrin distribution in malignant melanoma: association of the $\beta 3$ subunit with tumor progression. Cancer Res 50:6757–6764

Bowditch RD, Hariharan M, Tominna EF et al 1994 Identification of a novel integrin binding site in fibronectin. Differential utilization by $\beta 3$ integrins. J Biol Chem 269:10856–10863

Chan BMC, Kassner PD, Schiro JA, Byers HR, Kupper TS, Hemler ME 1992 Distinct cellular functions mediated by different VLA integrin α subunit cytoplasmic domains. Cell 68:1051–1060

Chen Y, Pitzenberger SM, Garsky VM, Lumma PK, Sanyal G, Baum J 1991 Proton NMR assignments and secondary structure of the snake venom protein echistatin. Biochemistry 30:11625–11636

Clements JM, Newham P, Shepherd M et al 1994 Identification of a key integrin-binding sequence in VCAM-1 homologous to the LDV active site in fibronectin. J Cell Sci 107:2127–2135

Elices MJ, Osborn L, Takada Y et al 1990 VCAM-1 on activated endothelium interacts with the leukocyte integrin VLA-4 at a site distinct from the VLA-4/fibronectin binding site. Cell 60:577–584

Frelinger AL, Cohen I, Plow EF et al 1990 Selective inhibition of integrin function by antibodies specific for ligand-occupied receptor conformers. J Biol Chem 265:6346–6352

Guan J-L, Hynes RO 1990 Lymphoid cells recognize an alternatively spliced segment of fibronectin via the integrin receptor $\alpha_4\beta_1$. Cell 60:53–61

Hemler ME, Elices MJ, Parker C, Takada Y 1990 Structure of the integrin VLA-4 and its cell–cell and cell–matrix adhesion functions. Immunol Rev 114:45–65

Humphries MJ 1990 The molecular basis and specificity of integrin–ligand interactions. J Cell Sci 97:585–592

Humphries MJ, Akiyama SK, Komoriya A, Olden K, Yamada KM 1986 Identification of an alternatively spliced site in human plasma fibronectin that mediates cell type-specific adhesion. J Cell Biol 103:2637–2647

Humphries MJ, Komoriya A, Akiyama SK, Olden K, Yamada KM 1987 Identification of two distinct regions of the type III connecting segment of human plasma fibronectin that promote cell type-specific adhesion. J Biol Chem 262:6886–6892

Humphries MJ, Doyle PM, Harris CJ 1994 Integrin antagonists as modulators of adhesion. Expert Opin Ther Pat 4:227–235

Hynes RO 1992 Integrins: versatility, modulation, and signaling in cell adhesion. Cell 69:11–25

Kloczewiak M, Timmons S, Lukas TJ, Hawiger J 1984 Platelet receptor recognition site on human fibrinogen. Synthesis and structure–function relationship of peptides corresponding to the carboxy-terminal segment of the γ-chain. Biochemistry 23:1767–1774

Komoriya A, Green LJ, Mervic M, Yamada SS, Yamada KM, Humphries MJ 1991 The minimal essential sequence for a major cell type-specific adhesion site (CS1) within the alternatively spliced type III connecting segment domain of fibronectin is leucine-aspartic acid-valine. J Biol Chem 266:15075–15079

Kraulis PJ 1991 MOLSCRIPT: a program to produce both detailed and schematic plots of protein structures. J Appl Crystallogr 24:946–950

Leahy DJ, Hendrickson WA, Aukhil I, Erickson HP 1992 Structure of a fibronectin type III domain from tenascin phased by MAD analysis of the selenomethionyl protein. Science 258:987–991

Li R, Nortamo P, Valmu L et al 1993 A peptide from ICAM-2 binds to the leukocyte integrin CD11a/CD18 and inhibits endothelial cell adhesion. J Biol Chem 268: 17513–17518

Logan D, Abu-Ghazaleh R, Blakemore W et al 1993 Structure of a major immunogenic site on foot-and-mouth disease virus. Nature 362:566–568

Main AL, Harvey TS, Baron M, Boyd J, Campbell ID 1992 The three-dimensional structure of the tenth type III module of fibronectin: an insight into RGD-mediated interactions. Cell 71:671–678

Makarem R, Newham O, Askari JA et al 1994 Competitive binding of vascular cell adhesion molecule-1 and the HepII/IIICS domain of fibronectin to the integrin $\alpha4\beta1$ J Biol Chem 269:4005–4011

Mould AP, Humphries MJ 1991 Identification of a novel recognition sequence for the integrin $\alpha4\beta1$ in the carboxy-terminal heparin-binding domain of fibronectin. EMBO (Eur Mol Biol Organ) J 10:4089–4095

Mould AP, Wheldon LA, Komoriya A, Wayner EA, Yamada KM, Humphries MJ 1990 Affinity chromatographic isolation of the melanoma adhesion receptor for the IIICS region of fibronectin and its identification as the integrin $\alpha4\beta1$. J Biol Chem 265:4020–4024

Mould AP, Komoriya A, Yamada KM, Humphries MJ 1991 The CS5 peptide is a second site in the IIICS region of fibronectin recognized by the integrin $\alpha4\beta1$. Inhibition of $\alpha4\beta1$ function by RGD peptide homologues. J Biol Chem 266: 3579–3585

Mould AP, Askari JA, Craig SE, Garratt AN, Clements J, Humphries MJ 1994 Cell adhesion and migration mediated by the integrin $\alpha4\beta1$: effect of fibronectin alternative splicing and the role of ligand affinity. J Biol Chem, in press

Needham LA, Van Dijk S, Pigott R et al 1994 Activation dependent and independent binding sites on vascular cell adhesion molecule-1. Cell Adhes Commun 2:87–99

Obara M, Kang MS, Yamada KM 1988 Site-directed mutagenesis of the cell-binding domain of human fibronectin: separable synergistic sites mediate adhesive function. Cell 53:649–657

Pierschbacher MD, Ruoslahti E 1984 Cell attachment activity of fibronectin can be duplicated by small synthetic fragments of the molecule. Nature 309:30–33

Santoro SA, Lawing WJ 1987 Competition for related but non identical binding sites on the glycoprotein IIb-IIIa complex by peptides derived from platelet adhesive proteins. Cell 48:867–873

Staatz WD, Fok KF, Zutter MM, Adams SP, Rodriguez BA, Santora SA 1991 Identification of a tetrapeptide recognition sequence for the $\alpha2\beta1$ integrin in collagen. J Biol Chem 266:7363–7367

Staunton DE, Dustin ML, Erickson HP, Springer TA 1990 The arrangement of the immunoglobulin-like domains of ICAM-1 and the binding sites for LFA-1 and rhinovirus. Cell 61:243–254

Vonderheide RH, Springer TA 1992 Lymphocyte adhesion through very late antigen 4: evidence for a novel binding site in the alternatively spliced domain of vascular cell adhesion molecule 1 and an additional $\alpha 4$ integrin counter-receptor on stimulated endothelium. J Exp Med 175:1433–1442
Wayner EA, Garcia-Pardo A, Humphries MJ, McDonald JA, Carter WG 1989 Identification and characterization of the T lymphocyte adhesion receptor for an alternative cell attachment domain CS-1 in plasma fibronectin. J Cell Biol 109:1321–1330
Weg VB, Williams TJ, Lobb RR, Nourshargh S 1993 A monoclonal antibody recognizing very late activation antigen-4 inhibits eosinophil accumulation *in vivo*. J Exp Med 177:561–566
Yamada KM 1991 Adhesive recognition sequences. J Biol Chem 266:12809–12812
Yednock TA, Cannon C, Fritz LC, Sanchez-Madrid F, Steinman L, Karin N 1992 Prevention of experimental autoimmune encephalomyelitis by antibodies against $\alpha 4\beta 1$ integrin. Nature 356:63–66
Zamarron C, Ginsberg MH, Plow EF 1991 A receptor-induced binding site in fibrinogen elicited by its interaction with platelet membrane glycoprotein IIb-IIIa. J Biol Chem 266:16193–16199

DISCUSSION

Hynes: You don't see any inhibition of adhesion to collagen by the peptide DGEA, so how do you explain Santoro's results (Staatz et al 1991), which look pretty convincing? Do you understand why yours are different?

Humphries: No.

Sonnenberg: I'm not convinced that the peptide used by Santoro is the only relevant peptide sequence.

Hynes: It may not be the *only* one, but Santoro's data that it was an $\alpha_2\beta_1$ inhibitor weren't bad.

Humphries: The concentrations of peptide that they used were very high. In order to see activity with a truncated peptide, I believe they needed concentrations of 5–6 mM.

Hynes: Do you see the I-domain of $\alpha_2\beta_1$ binding to laminin as well? This is also a laminin receptor in some situations.

Humphries: No, we didn't see any binding with the α_2 I-domain.

Sonnenberg: Can you block $\alpha_1\beta_1$ binding to collagen with the I-domain of α_2?

Humphries: We've not tried that.

Rothlein: Did you look for the integrin-binding sequence in domain III of ICAM-1, which binds to $\alpha_M\beta_2$ (Mac-1)?

Humphries: Yes; there doesn't appear to be a homologous sequence. I don't know what that says about ICAM-1–$\alpha_M\beta_2$ binding.

Rothlein: It seems pretty clear that when you have an I-domain, antibodies that block cell adhesion bind to the I-domain; antibodies that don't bind to the I-domain are not effective blockers of cell adhesion, yet most anti-β-subunit antibodies block effectively. What's the relationship between the α and the β subunits?

Humphries: That's a central question. We desperately need to understand how the ligand-binding pocket is formed. The information could also impact upon activation, drug development and all kinds of phenomena. I should add that there are examples of antibodies that don't bind to the I-domain blocking function. Danny Tuckwell has found that HAS-6, one of Fiona Watt's anti-α_2 antibodies, is a weak blocker of adhesion and doesn't map to the I-domain (Tenchini et al 1993).

Of course, we should remember that the I-domain isn't found in every integrin and therefore can't explain all ligand-binding events.

I should also add that Klaus Kuhn's lab have reported a quite interesting hypothesis about how type IV collagen is recognized by $\alpha_1\beta_1$ (Eble et al 1993). They're lucky that they can actually fragment type IV collagen and retain a triple helical structure that is held together by disulphide bonds. By chemical modification of amino acids, they've shown that there is an arginine residue and an aspartate residue that are crucial for that binding. However, these two residues are not on the same chain in the trimer, but are actually found on adjacent chains. They think that there may well actually be an RGD type of recognition of the type IV collagen mediated by a basic and an acidic group on different chains.

Sonnenberg: But your isolated I-domain doesn't bind to collagen IV?

Humphries: It binds, but weakly. It also binds weakly to type VI, which is very interesting.

Sonnenberg: How do you explain that? Are there two binding sites?

Humphries: I think there's more than one binding site and that, depending on the ligand, each is used to a different extent.

Elices: What is the relationship between the solid-phase assays you do (in which you use the purified receptor) and the receptor that is expressed on cells?

Humphries: Fundamentally they are the same. The binding we see in the solid-phase assays is specific and occurs under conditions that support adhesion to authentic ligands. In cell-based assays the environment of the integrin might be modulated to produce receptors with different activities. This is where cell adhesion experiments can actually give different results. For example, differences in receptor presentation and in intracellular activation or clustering of the integrin can give you apparent differences in cation sensitivity or sensitivity of the adhesion to inhibitors.

Shaltiel: Can you tell us more about the effect of the cations on the efficacy of the different peptides?

Humphries: It's dependent on the integrin. In a way it's quite reassuring to see a different pattern for the α_2 I-domain and for α_4. Paul Mould (unpublished results) has done similar experiments with $\alpha_5\beta_1$, using the 80 kDa fragment of fibronectin as a biotinated ligand, and there you see a very different pattern to α_4. Whereas Ca^{2+} is slightly better than Mg^{2+} for α_4 ligands, with the 80 kDa ligand it's the other way round. I don't know whether or not there is any difference in cation specificity for different ligands for the same integrin. Current evidence suggests the cation pattern is the same.

Hogg: Martin Hemler has shown differences in the divalent cation dependency pattern for $\alpha_4\beta_1$ binding to fibronectin (Mg^{2+} only) versus VCAM-1 (Mg^{2+} and Ca^{2+}; Matsumoto & Hemler 1993).

Humphries: With isolated integrin?

Hogg: No, on the cell surface.

Humphries: You can get misleading results by concentrating just on cell adhesion assays because cation status is only one factor affecting net affinity of ligand binding; the inherent affinity of the ligand, the number of receptors and the state of receptor activation must also be taken into account.

Pober: When you did experiments studying cell attachment to the denatured collagen fragments and showed that it was an α_5-mediated effect, is that because of direct recognition of these gelatin fragments by α_5, or are the cells bringing fibronectin with them that binds to the gelatin, producing attachment through a fibronectin bridge?

Humphries: They are making fibronectin. It's well known that chondrocytic cells make huge amounts of fibronectin when they're released from the cartilage matrix. By immunofluorescence, Danny Tuckwell has observed fibronectin and α_5 in focal contacts on denatured collagen.

Pober: So it's not direct recognition of collagen by α_5 *per se*?

Humphries: No, it's a bridge. I should just add that the CB10 fragment, which is the most adhesive, is already characterized as being the major fibronectin binding site in type II collagen.

Pober: You pointed out the alternative splicing of fibronectin to either include or exclude a ligand, and you made the same comment about VCAM. I just want to throw out something for consideration about VCAM splicing. There does seem to be a profusion of molecular VCAM species in mice and rabbits. However, it's not obvious that there is alternative splicing of VCAM in humans. There is a minor species of VCAM mRNA that is present lacking exon 4, but it's always a minor species. No one has been able to find a human cell type in which there is ever more than a tiny fraction of the exon 4-deleted message, and no one has demonstrated this form of protein in a cell other than a transfectant made with the cDNA lacking exon 4.

Humphries: But it may be that those different domains in the full length molecule mediate different functions. There is some evidence for temporal differences in the use of domains I and IV during cell adhesion.

Sonnenberg: To see collagen binding you need quite high concentrations of manganese: why is this?

Humphries: You don't actually need much, but a plateau isn't reached until you use concentrations higher than 10 mM.

Sonnenberg: Is the metal binding site in the I-domain that weak?

Humphries: No, this is true for purified integrins as well. You don't normally add that amount because of complications with the assay; you certainly still get effects at sub-millimolar levels.

Elices: You mentioned that the binding of $\alpha_2\beta_1$ to collagen is conformation dependent. However, in the earlier peptide experiments, you showed that different peptides were fairly effective at inhibiting $\alpha_4\beta_1$ interaction with its ligands. How do you reconcile the apparent inconsistency that in some cases conformation seems to be important (for example, in $\alpha_2\beta_1$ binding to collagen) and in other cases it isn't.

Humphries: I think it's important in all cases. The difference between those two systems is that for the α_4 ligands the active site is in a linear peptide; the fact that you see inhibition of one by the other is because they use the same active site. With collagens, my guess is that there isn't a linear peptide that can explain that binding, so you need contributions from more than one chain and this wouldn't be possible in a fragmented molecule.

Garrod: Is anything known about α_1 binding to collagen? Are there any known parallels?

Humphries: From adhesion assays, we know that the specificity of ligand binding to the two integrins is slightly different. Overall, α_1 tends to prefer type IV collagen. $\alpha_2\beta_1$ does bind to type IV and type VI collagen, but in I-domain binding you see differences: that's why I think there may be differential use of different sites in the subunit. It will be important to test the α_1 I-domain in parallel with α_2.

References

Eble JA, Golbik R, Mann K, Kuhn K 1993 The $\alpha_1\beta_1$ integrin recognition site of the basement membrane collagen molecule [α_1(IV)$_2\alpha_2$(IV)]. EMBO (Eur Mol Biol Organ) J 12:4795–4802

Matsumoto A, Hemler ME 1993 Multiple activation states of VLA-4: mechanistic differences between adhesion to CS1/fibronectin and to vascular cell adhesion molecule. J Biol Chem 268:228–234

Staatz WD, Fok KF, Zutter MM, Adams SP, Rodriguez BA, Santoro SA 1991 Identification of a tetrapeptide recognition sequence for the $\alpha_2\beta_1$ integrin in collagen. J Biol Chem 266:7363–7367

Tenchini ML, Adams JC, Gilberty C et al 1993 Evidence against a major role for integrins in calcium-dependent intercellular adhesion of epidermal keratinocytes. Cell Adhes Commun 1:35–66

General discussion III

Binding sites on the integrin $\alpha_L\beta_2$ (LFA-1) for its ligand ICAM-1

Hogg: I am going to describe work carried out by Anna Randi and Paula Stanley in my laboratory. The integrin $\alpha_L\beta_2$ (also known as lymphocyte function-associated antigen 1 [LFA-1]) has specificity for intercellular adhesion molecule 1 (ICAM-1), but it has been notably difficult to demonstrate binding of ICAM-1 to intact $\alpha_L\beta_2$ in solution. In order to identify binding sites on $\alpha_L\beta_2$ for ICAM-1, we have taken the approach of translating *in vitro* regions of the α_L subunit to obtain a 'nested' series of protein fragments with domains deleted from the N-terminus of the subunit (Fig. 1). We have then used a chimeric form of ICAM-1 with an IgG1 Fc 'tail' to precipitate these ^{35}S-methionine-labelled protein fragments. The minimum fragment with which we could obtain precipitation with ICAM-1 corresponded to domains V and VI, which are two of the three putative divalent cation-binding domains within the α subunit (i.e. domains V–VII) (Stanley et al 1994).

This result indicated that a binding site for ICAM-1 is contained within domain V and VI of the α_L subunit. In order more precisely to pinpoint the ICAM-1 binding site within this region, we next made a series of 20mer peptides covering domains IV through VII and tested their ability to inhibit T cell binding to ICAM-1 via $\alpha_L\beta_2$. The two peptides which inhibited the assay are located downstream of the putative divalent-cation binding site of domain V and overlapping the metal binding sequence of domain VI. Because domains V–VII have been speculated to resemble the classic helix-loop-helix motif of the Ca^{2+}-binding EF hand, we have modelled domain V and VI on the basis of the solved structure of the calmodulin EF hands (Stanley et al 1994). When the inhibitory peptides are positioned on the model, the two ICAM-1 binding sites lie adjacent to one another along one rim of the model and appear to form a single binding site.

When the I-domain was eliminated from the $\alpha_L\beta_2$-translated α-subunit fragments, however, there was a large decrease in the ability of ICAM-1 to precipitate $\alpha_L\beta_2$ fragments, suggesting that the I-domain is involved in binding of ICAM-1. To investigate its role in ICAM-1 binding, we expressed recombinant isolated I-domain as a fusion protein with the Fc portion of IgG1 (I-Fc) and tested a panel of anti-$\alpha_L\beta_2$ monoclonal antibodies for their ability to interact with it. The majority of the previously characterized anti-$\alpha_L\beta_2$ antibodies (18 out of 20) recognized epitopes within the I-domain (Randi & Hogg 1994). This may reflect a structural feature, e.g. this region could be more exposed and, therefore, more immunogenic than other parts of the molecule. Many of these

I-domain monoclonal antibodies block the $\alpha_L\beta_2$/ICAM-1 interaction, providing further evidence for the involvement of this domain in ligand binding. The domain could regulate binding by controlling the exposure of a cryptic binding site, or could function as a binding site itself. To test these options, we investigated the effect of recombinant $\alpha_L\beta_2$ I-domain (I-Fc) on the binding of T cells to ICAM-1 in an assay which is $\alpha_L\beta_2$/ICAM-1 dependent. I-Fc inhibited T cell binding to ICAM-1 (Fig. 2), whereas the control protein CD14-Fc had no effect on the assay; I-Fc also bound directly to immobilized recombinant ICAM-1 (Randi & Hogg 1994).

Therefore, there are at least two ICAM-1 binding sites on the α_L subunit: one in the I-domain and one in the divalent cation-binding domains V/VI. In addition, there is evidence for an RGD binding site at the N-terminus of the β_3 subunit, within an area that is highly homologous among all β-subunits (Loftus et al 1990). This suggests that the β_2 subunit may also contribute to ligand binding. How all of these sites actually work together to bind the ligand is the next question. They may act in concert, as a sort of 'maxi' ligand binding pocket, or in sequence, with the binding to one site promoting interaction with a second site. With regard to this second possibility, the notion of sequential activation arises, with the interaction of ICAM-1 with one binding site on $\alpha_L\beta_2$ leading to the exposure of a second site, resulting in more stable binding of the $\alpha_L\beta_2$/ICAM-1 receptor pair (Cabañas & Hogg 1993).

Rothlein: A lot of those antibodies against $\alpha_L\beta_2$ were screened functionally. This may be why they are directed against the I-domain, not because the I-domain is immunodominant. Have you tried blocking through the I-domain with peptides?

Hogg: We are in the process of doing that sort of experiment.

Rothlein: Do your results indicate that the I-domain has stronger binding than domains V and VI?

Hogg: Potentially, yes, but peptides from the domain V/VI region also block a T lymphocyte assay which measures $\alpha_L\beta_2$/ICAM-1 binding.

Rothlein: Do they block the purified α-subunit if you do immunoprecipitation?

Hogg: The peptides each block the T cell assay but a peptide covering the domain V site also blocks in solution. This suggests either that there is a sequential interaction between the binding sites (so if you block at any one point you block the entire interaction) or that they each critically contribute energy to the binding site (so that if you diminish one site of interaction you destabilize the whole site).

Sonnenberg: If all these proteins were made by recombinant methods in bacteria, it means that they are not glycosylated.

Hogg: The α_L protein fragments are made *in vitro* in the standard reticulocyte lysate system, and the I-domain fragment is obtained from COS cells. However, the $\alpha_L\beta_2$ I-domain protein doesn't have any *N*-linked glycosylation, nor does the domain V/VI region.

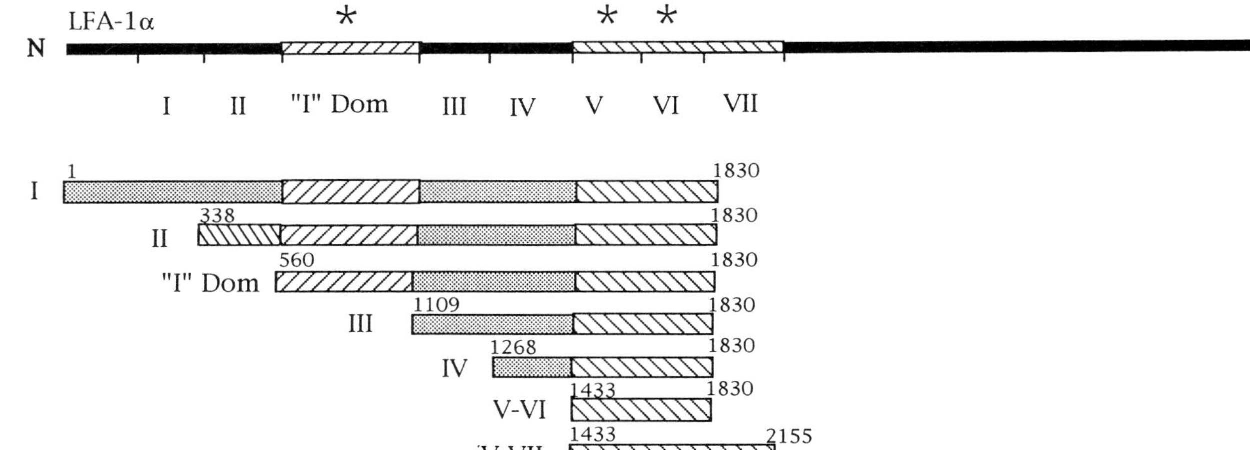

FIG. 1. The α_L (LFA-1 α-subunit) deletion series used to show binding of ICAM-1 to domains V and VI. Domains are successively deleted from the N-terminus and fragments terminate at the beginning of domain VII. The positions of ICAM-1 binding sites located on the α_L subunit are indicated by asterisks.

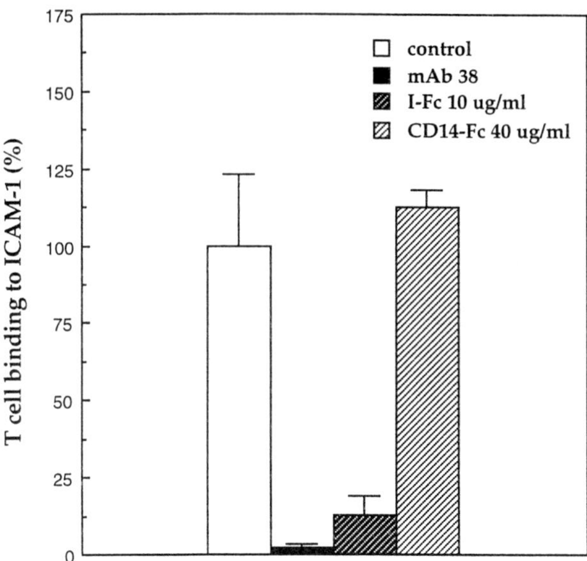

FIG. 2. T cell binding to recombinant ICAM-1-Fc. T cell binding is expressed as percentage of control binding. The inhibition of binding by a known anti-$\alpha_L\beta_2$ (LFA-1) monoclonal antibody (mAb 38) and by recombinant chimeric $\alpha_L\beta_2$ I-domain (I-Fc) is shown. The control chimeric protein CD14-Fc does not affect binding. Bars represent mean $\pm$ standard deviation of triplicates of one experiment representative of five.

Shaltiel: Is there synergism in the inhibition by the two sites?

Hogg: We haven't yet done those experiments!

Elices: Both Martin Humphries and Nancy Hogg have mentioned the importance of cation binding sites—what is the involvement of these in the binding event?

Hogg: We found that binding to the domain V and VI site, at least, was not divalent-cation dependent (Stanley et al 1994). This was unexpected and raises the question of what the role of divalent cation action might be when bound to these domains. There are lots of examples in other proteins of divalent cations acting as modulators of tertiary or quaternary structure. So it is possible that divalent cations don't affect the local secondary structure, but they might affect the relationship between the α and β subunit of an integrin. This is just a speculation for which there is no evidence at present.

Pober: Given the fact that the domain V/VI peptides do not appear to have a particular affinity for ICAM, could they be binding to $\alpha_L\beta_2$ and be perturbing its structure by competing for an internal interaction?

Hogg: That is an interesting idea. However, the fact that at least the domain V peptide can interfere with a truncated α-subunit fragment in solution does indicate direct binding to ICAM-1 but we obviously cannot rule out the possibility that they are perturbing some other part of the structure.

Humphries: I believe the cation sites are very important, but my feeling is that they probably do more than one job. There's a lot of evidence for a role for cations in the activation of integrins as well as some very good evidence for a role in ligand binding. In addition, I think they are critical for α,β orientation. Hence, in experiments where you mutate those sites, the interpretation of the results may be complicated.

Elices: I take it from Nancy Hogg's comments that divalent cations might not be involved in the binding event *per se*. Rather, they may participate in sustaining the 3D structure.

Hogg: Or, potentially, revealing the binding site.

Humphries: You can't tell yet, there's not enough information.

References

Cabañas C, Hogg N 1993 Ligand intercellular adhesion molecule-1 has a necessary role in activation of integrin lymphocyte function-associated molecule-1. Proc Natl Acad Sci USA 90:5838–5842

Loftus JC, O'Toole TE, Plow EF, Glass A, Frelinger AL, Ginsberg MH 1990 A β_3 integrin mutation abolishes ligand binding and alters divalent cation-dependent conformation. Science 249:915–918

Randi AM, Hogg N 1994 I-domain of β_2 integrin lymphocyte function-associated antigen-1 contains a binding site for ligand intercellular adhesion molecule-1. J Biol Chem 269:12395–12398

Stanley P, Bates PA, Harvey J, Bennett RI, Hogg N 1994 Integrin LFA-1 α subunit contains an ICAM-1 binding site in domains V and VI. EMBO (Eur Mol Biol Organ) J 13:1790–1798

Treatment of inflammatory diseases with a monoclonal antibody to intercellular adhesion molecule 1

Robert Rothlein and Judith R. Jaeger

Boehringer Ingelheim Pharmaceuticals Inc., 900 Ridgefield Road, PO Box 368, Ridgefield, CT 06877, USA

Abstract. A mouse monoclonal antibody to intercellular adhesion molecule 1 (ICAM-1) has been evaluated in multiple animal models of inflammation as well as in the clinic. Anti-ICAM-1 has been found to protect against allograft rejection and ischaemia/reperfusion injury in non-human primates and rabbits. In open-label phase I–II studies, anti-ICAM-1 appears to have prolonged kidney allograft survival when used as induction therapy in conjunction with traditional triple immunosuppressive therapy. Anti-ICAM-1 has shown beneficial effects in the treatment of rheumatoid arthritis when given for five days. Most patients receiving anti-ICAM-1 made antibodies to the mouse immunoglobulin.

1995 Cell adhesion and human disease. Wiley, Chichester (Ciba Foundation Symposium 189) p 200–211

It is clear that leukocyte–leukocyte and leukocyte–target cell adhesion are necessary events for the host defence system to function normally in inflammation. Specifically, leukocytes must first attach to the endothelial cells which line blood vessels, prior to migrating from the circulation to sites of inflammation (Perry & Granger 1991, Argenbright et al 1991). Also, lymphocytes must adhere to antigen-presenting cells in order for specific immunological processes such as antibody production or the generation and elaboration of antigen-specific T cells to occur (Lipsky & Rosenthal 1975). Lastly, for leukocytes to act as effectors to mediate lysis of target cells, they must first attach to these target cells. In the latter case, attachment both co-stimulates and/or triggers the effector cell to secrete its lytic mediators as well as presumably allowing the lytic mediators secreted by the leukocytes to achieve high local concentrations at the leukocyte–target cell junction, thus facilitating the destruction of the target cell (Wacholtz et al 1989, Van Seventer et al 1991).

ICAM-1

Over the past 7–8 years, we have been evaluating the anti-inflammatory activity of a monoclonal antibody to intercellular adhesion molecule 1 (ICAM-1) as a potential therapeutic agent in the treatment of inflammatory diseases.

ICAM-1 is a member of the immunoglobulin supergene family with five Ig-like domains, a single membrane-spanning region and a short cytoplasmic tail (Staunton et al 1988, Simmons et al 1988). Except for a low basal expression on endothelial cells, there is little, if any, ICAM-1 expressed on cells in unperturbed tissue. However, ICAM-1 is inducible *in vitro* on many cell types, including haemopoietic cells, endothelial cells and fibroblasts, with cytokines such as interleukin (IL)-1α, IL-1β, tumour necrosis factor (TNF)-α, TNF-β, and/or interferon-γ depending on the cell type (Dustin et al 1986, 1988, Pober et al 1986, Frohman et al 1989). Furthermore, ICAM-1 is expressed *in vivo* at sites of inflammation. As mentioned previously, there is a low constitutive expression of ICAM-1 on venular endothelial cells; however, its expression is markedly increased at inflammatory sites (Dustin et al 1986). Furthermore, there is increased ICAM-1 expression on many other cell types including keratinocytes in inflammatory skin lesions (Vejlsgaard et al 1989, Griffiths & Nickoloff 1989), transplanted liver bile duct and perivenular hepatocytes during rejection (Adams et al 1989) and in native livers with primary biliary cirrhosis (Adams et al 1991) at sites of inflammation. ICAM-1 expression is also increased on endothelium in the human brain surrounding multiple sclerosis plaques (Sobel et al 1990, Frohman et al 1989, Raine et al 1990), transplanted kidney glomeruli and tubules during rejection (Cosimi et al 1990), and lung epithelial cells following antigen provocation (Wegner et al 1990). Increased ICAM-1 expression is found on melanoma cells following metastasis (Matsui et al 1987, Natali et al 1990).

Initially, when ICAM-1 was identified as the ligand of $\alpha_L\beta_2$ (also known as LFA-1), it was thought to be functional only in lymphocyte-mediated processes. Smith et al (1988), using anti-ICAM-1 in a neutrophil/endothelial cell adhesion assay, found that ICAM-1 also inhibited neutrophil adhesion to and migration through endothelium. They also found that there was both an $\alpha_L\beta_2$ component and $\alpha_M\beta_2$ (Mac-1) component to the neutrophil–ICAM-1 adhesion depending on the activation state of the neutrophil, suggesting that both $\alpha_M\beta_2$ and $\alpha_L\beta_2$ bind to ICAM-1 (Smith et al 1989). This finding was later confirmed with a direct binding assay showing that $\alpha_L\beta_2$ recognizes domain I and $\alpha_M\beta_2$ recognizes domain III of ICAM-1 (Staunton et al 1990, Diamond et al 1990). In summary, ICAM-1 mediates both lymphocyte and granulocyte adhesion processes as shown by functional *in vitro* assays such as mixed lymphocyte reactions (Boyd et al 1988), antigen-induced proliferation (Dougherty et al 1988), cytotoxic T cell activity (Mentzer et al 1988, Makgoba et al 1989) and granulocyte and lymphocyte attachment to endothelium (Smith et al 1988, 1989, Dustin & Springer 1988) that are inhibited by anti-ICAM-1 monoclonal antibody.

Encouraged by the above *in vitro* data, we evaluated the *in vivo* activity of antibodies to ICAM-1 in animal models of neutrophil- and/or lymphocyte-mediated inflammatory disease. One of the earliest experiments to be performed was that by Barton et al (1989) who demonstrated that anti-ICAM-1 inhibited neutrophil influx into rabbit lungs following systemic activation with phorbol esters, providing *in vivo* evidence that anti-ICAM-1 monoclonal antibody did indeed block neutrophil function. Subsequently, it was shown that anti-ICAM-1 monoclonal antibody also mitigated eosinophil influx and airway hyper-responsiveness in a non-human primate model of antigen-induced airway hyper-responsiveness (Wegner et al 1990)—a model that has both non-specific and specific inflammatory components. This suggested a potential clinical role for the use of anti-ICAM-1 antibodies in both acute and chronic pulmonary diseases.

Subsequently, ICAM-1 antagonism was tested in a model of immunologic kidney allograft rejection. When anti-ICAM-1 monoclonal antibody was given as the sole form of immunosuppressive therapy for 12 days only, starting 2 days prior to transplant, it increased the time to rejection of allogeneic kidneys in non-human primate recipients. The average time of allograft survival more than doubled from an average of less than 10 days in the untreated animals to more than 22 days in the treated animals (Cosimi et al 1990). This provided *in vivo* evidence of the ability of an antibody to ICAM-1 to inhibit lymphocyte function.

A second set of experiments on this same model system showed that anti-ICAM-1 monoclonal antibodies also reversed acute allograft rejection episodes. In these experiments, monkeys received kidney transplants with cyclosporin A (CsA) as the sole immunosuppressive therapeutic agent. CsA doses were then tapered until sub-therapeutic levels were achieved. This resulted in an acute rejection episode, detected by a rise in serum creatinine. At this point, anti-ICAM-1 was administered daily for 10 days while the sub-therapeutic dose of CsA was maintained. These experiments revealed that anti-ICAM-1 normalized the rise in serum creatinine in all animals tested and markedly prolonged kidney allograft survival time (Cosimi et al 1990). These data thus suggested a role for an ICAM-1 antagonist in mitigating specific immunological processes such as those involved in allogeneic solid-organ transplantation.

Another model of disease in which we found anti-ICAM-1 to be effective in mitigating the inflammatory response is a rabbit spinal cord ischaemia/reperfusion model. This model, thought to be representative of stroke, tests the effects of an adhesion antagonist on neutrophil-mediated damage. We found that antibodies to either β_2 integrin (CD18) or ICAM-1 allowed the rabbit to be subjected to longer ischaemic times without any apparent subsequent deleterious effects (Clark et al 1991a,b). More recently, we reported that anti-ICAM-1 treatment allowed rabbits to withstand more microthrombi injected into the carotid artery before sustaining permanent neurological damage than rabbits treated with saline (Bowes et al 1993).

Finally, it was recently reported that anti-ICAM-1 treatment given up to three hours post-burn was able to inhibit the extension of full-thickness burn into zones of stasis when compared with saline treatment in rabbits (Mileski et al 1992).

Anti-ICAM-1 in the clinic

The findings thus far described support a role for the use of anti-ICAM-1 in treating a number of diseases. The specific anti-ICAM-1 selected for clinical trials is an IgG2a which binds to domain II of ICAM-1. This antibody, called BIRR1 (also called R6.5), blocks both lymphocyte and neutrophil adhesion (Smith et al 1988, Diamond et al 1990). BIRR1 does not inhibit bactericidal or phagocytic functions of granulocytes, as would an antagonist of β_2 integrin (Mileski et al 1993).

The first clinical areas selected for study with BIRR1 as a proof of concept have been solid organ transplantation and rheumatoid arthritis. Phase I–II studies in stroke and burn are also being planned. The unavailability of a less immunogenic therapeutic agent, such as a 'humanized' anti-ICAM-1 or a low molecular weight compound, limit the full exploration of the therapeutic potential of ICAM-1 antagonism at this time, but early evidence suggests that even the murine antibody is effective and relatively safe.

Renal transplant

The renal transplant program was the first clinical program to start. It is currently in Phase II with a double-blind, placebo-controlled study ongoing. A Phase I–II trial in liver transplantation has also been completed.

The phase I–II renal transplantation study was performed at the Massachusetts General Hospital by Dr C. B. Cosimi. This study involved cadaveric renal allograft recipients at high risk for post-transplant complications such as delayed graft function (DGF) and acute rejection. These patients were given BIRR1 prophylactically as an add-on medication to conventional triple immunosuppressive therapy (azathioprine, steroids and CsA). The objectives of this study were to evaluate the safety and pharmacokinetics of BIRR1 and to attempt to get a clinical read-out on the ability of BIRR1 to reduce the incidence of DGF and acute rejection (Haug et al 1993).

It was originally intended that only 10 patients, five of each on two dosing regimens, would be studied. These dosing regimens were direct extrapolations from those which were efficacious in the cynomolgus monkeys, as well as the investigator's experience with OKT3 (another mouse IgG2a monoclonal antibody). It was anticipated that both of these regimens would provide BIRR1 levels in a potentially therapeutic range.

It turned out that the original dosing projections were far too low, and the treatment period (two weeks) was impractically long. Thus, in order to fulfil the objective of the study (i.e. to find two dosing regimens which achieved the pharmacokinetic target), a total of five dosing regimens were studied in 18 patients. The treatments ranged from a total of 150 mg of antibody over a 14 day period to a total of 560 mg of antibody over a 6 day period. Of the five dosing regimens studied, the two highest regimens consistently met the pharmacokinetic target. Both were 6 day regimens: 160 mg pre-transplant loading, followed by 40 mg/day × 5 days (abbreviated as 160/40 × 5), and a 160/80 × 5 regimen, for total doses of 360 mg and 560 mg, respectively.

The reasons for the increased dosing requirement for BIRR1 in renal allograft recipients when compared with animal studies may have to do with any one (or a combination) of the following:

(i) Expression of ICAM-1 in a chronically ill human versus a healthy monkey. In this situation ICAM-1, which is normally present only in small amounts, is markedly up-regulated in the presence of inflammation. This could cause an 'ICAM-1 sink,' requiring more anti-ICAM-1 in order to saturate all the ICAM-1 sites. We will learn more about this as we investigate BIRR1 in different diseases.
(ii) Higher avidity of BIRR1 for human ICAM-1 than non-human primate ICAM-1 with a corresponding slower dissociation rate.
(iii) Other factors not yet determined.

As the study was designed originally, efficacy comparisons could only be made with a historical control. As the study evolved, however, 18 patients were enrolled who received five different regimens. Eleven patients received anti-ICAM-1 regimens which achieved the pharmacokinetic target; seven did not. The outcome of these two groups can be compared.

The incidence of both acute rejection (followed for three months) and DGF (a shorter-term endpoint evaluable by the end of the first post-transplant week) were lower in the 11 patients with adequate BIRR1 levels. In fact, of the seven patients not achieving pharmacokinetic target levels of BIRR1, all had at least one rejection episode within the first three months post-transplant while only three of 11 patients achieving target levels of BIRR1 had a rejection episode within the first three months post-transplantation. In terms of DGF, anti-ICAM-1 also seemed to be protective since 6/7 of those patients with sub-therapeutic levels of BIRR1 had DGF while 6/11 of those patients achieving therapeutic levels had DGF.

Sixteen of the 18 patients developed an immune response to BIRR1 (human anti-mouse antibody [HAMA]), which is to be expected. A high incidence of BIRR1 anti-idiotype antibodies was found, but only two of 10 HAMA responders tested had antibodies which cross-reacted with OKT3, another mouse IgG2a monoclonal antibody. The above data suggest that anti-ICAM-1, as predicted from *in vitro* and *in vivo* models, attenuates both lymphocyte- and granulocyte-mediated reactions in humans. This is currently being evaluated in controlled clinical trials.

Rheumatoid arthritis

Anti-ICAM-1 is also being evaluated in an investigator's new drug study in an open-label, dose-escalating phase I–II study in patients with severe rheumatoid arthritis by Dr Peter Lipsky (University of Texas, Southwestern Medical Center, Dallas) (Kavanaugh et al 1992a,b). Although primarily a pharmacokinetic and safety study in its focus, efficacy is also assessed using the Paulus criteria described below. The population under study includes patients with active disease of greater than four years (average 18 years) duration who are functional class I–III and who have failed at least two disease-modifying drugs (average four drugs). Patients are allowed to be on non-steroidal anti-inflammatory drugs or $\leqslant 10$ mg prednisone/day during active treatment with BIRR1. Twenty-three patients have completed a full 5 day course of treatment with dosing regimens ranging from 60/20 mg × 4 d to 240/80 mg × 4 d with the majority of the patients receiving a regimen of 120/40 mg × 4 d. In all of the dosing regimens studied there have been no serious adverse events. Both the 60/20 mg × 4 d and 120/40 mg × 4 d regimens were well-tolerated while those patients receiving the 240 mg loading dose had a mild to moderate 'first dose reaction.' The pharmacokinetic target was achieved in all patients receiving the 120/40 mg × 4 d regimen.

Efficacy was assessed in this study using the modified Paulus criteria, which examine the following six parameters: joint swelling, joint tenderness, morning stiffness, patient/physician global assessment and erythrocyte sedimentation rate (ESR). In order to be classified as a 'responder' a patient must have a $\geqslant 20\%$ reduction in four out of six parameters.

Results to date reveal that of the 23 patients receiving full treatment, 13 were considered responders with nine showing a response up to 60 d post-treatment. Responses appeared to wane between 60–90 d post-treatment.

Thus, in severe rheumatoid arthritis, BIRR1 has been tested at several dosing regimens. The 120/40 mg × 4 d dosing regimen (280 mg total) is well tolerated, achieves the pharmacokinetic target, and has resulted in a clinical response in more than half the patients. Furthermore, during treatment, T cell counts in the blood were elevated but rapidly returned to normal levels upon cessation of therapy.

To summarize the clinical experience to date, in open-label studies, 18 kidney and 17 liver transplant recipients and more than 23 patients with rheumatoid arthritis have received BIRR1 using various dosing regimens. In each indication, a different dosing regimen has been required to achieve the pharmacokinetic target, suggesting that these diseases have different amounts of ICAM-1 expression, cICAM-1, and/or other factors that may influence BIRR1 serum levels. BIRR1 has been well tolerated at the dosing regimens which achieve the pharmacokinetic target. Safety and efficacy are being objectively assessed in an ongoing placebo-controlled, double-blind, renal transplant study involving 200 patients.

In conclusion, the inhibition of β_2 integrin/ICAM-1 interactions inhibits leukocyte trafficking, antigen presentation and leukocyte-mediated cytolysis. It is unclear which of the three processes inhibited by anti-ICAM-1 (summarized in this chapter) is most responsible for the dramatic reduction in the inflammatory responses. It will probably be a combination of all three. Regardless, it is clear that antagonists of these interactions offer an exciting new approach to prevent inflammatory responses associated with transplantation, autoimmunity and acute inflammation.

Acknowledgements

We would like to thank all of our many preclinical and clinical collaborators who have contributed to our understanding of ICAM-1 and identified clinical situations where anti-ICAM-1 therapy may be beneficial.

References

Adams DH, Hubscher SG, Shaw J, Rothlein R, Neuberger JM 1989 Intercellular adhesion molecule 1 on liver allografts during rejection. Lancet 2:1122–1125

Adams DH, Hubscher SG, Shaw et al 1991 Increased expression of intercellular adhesion molecule 1 on bile ducts in primary biliary cirrhosis and primary sclerosing cholangitis. Hepatology 14:426–431

Argenbright LW, Letts LG, Rothlein R 1991 Monoclonal antibodies to the leukocyte membrane CD18 glycoprotein complex and to intercellular adhesion molecule-1 inhibit leukocyte-endothelial adhesion in rabbits. J Leukocyte Biol 49:253–257

Barton RW, Rothlein R, Ksiazek J, Kennedy C 1989 The effect of anti-intercellular adhesion molecule-1 on phorbolester-induced rabbit lung inflammation. J Immunol 143:1278–1282

Bowes MP, Zivin JA, Rothlein R 1993 Monoclonal antibody to the ICAM-1 adhesion site reduces neurological damage in a rabbit cerebral embolism stroke model. Exp Neurol 119:215–219

Boyd AW, Wawryk SO, Burns GF, Fecondo JV 1988 Intercellular adhesion molecule 1 (ICAM-1) has a central role in cell–cell contact–mediated immune mechanisms. Proc Natl Acad Sci USA 85:3095–3099

Clark WM, Madden KP, Rothlein R, Zivin JA 1991a Reduction of central nervous system ischemic injury in rabbits using leukocyte adhesion antibody treatment. Stroke 22:877–883

Clark WM, Madden KP, Rothlein R, Zivin JA 1991b Reduction of central nervous system ischemic injury by monoclonal antibody to intercellular adhesion molecule. J Neurosurg 75:623–627

Cosimi AB, Conti D, Delmonico FL et al 1990 In vivo effects of monoclonal antibody to ICAM-1 (CD54) in nonhuman primates with renal allografts. J Immunol 144: 4604–4612

Diamond MS, Stauton DE, de Fougerolles AR et al 1990 ICAM-1 (CD54): a counter-receptor for Mac-1 (CD11b/CD18). J Cell Biol 111:3129–3139
Dougherty GJ, Murdoch S, Hogg N 1988 The function of human intercellular adhesion molecule-1 (ICAM-1) in the generation of an immune response. Eur J Immunol 18:35–39
Dustin ML, Springer TA 1988 Lymphocyte function associated antigen-1 (LFA-1) interaction with intercellular adhesion molecule-1 (ICAM-1) is one of at least three mechanisms for lymphocyte adhesion to cultured endothelial cells. J Cell Biol 107:321–331
Dustin ML, Rothlein R, Bhan AK, Dinarello CA, Springer TA 1986 Induction by IL-1 and interferon, tissue distribution, biochemistry, and function of a natural adherence molecule (ICAM-1). J Immunol 137:245–254
Dustin ML, Singer KH, Tuck DT, Springer TA 1988 Adhesion of T lymphoblasts to epidermal keratinocytes is regulated by interferon gamma and is mediated by intercellular adhesion molecule-1 (ICAM-1). J Exp Med 167:1323–1340
Frohman EM, Frohman TC, Dustin ML et al 1989 Induction of ICAM-1 expression on human fetal astrocytes by interferon-gamma, tumor necrosis factor-alpha and interleukin-1: relevance to intracerebral antigen presentation. J Neuroimmunol 23:117–124.
Griffiths CEM, Nickoloff BJ 1989 Keratinocyte intercellular adhesion molecule-1 (ICAM-1) expression precedes dermal T lymphocytic infiltration in allergic contact dermatitis (Rhus dermatitis). Am J Pathol 135:1045–1053
Haug CE, Colvin RB, Delmonico FL et al 1993 A phase I trial of immunosuppression with anti-ICAM-1 (CD54) mAb in renal allograft recipients. Transplantation 55:766–773
Kavanaugh A, Nichols LA, Lipsky PE 1992a Treatment of refractory rheumatoid arthritis with an anti-CD54 (intercellular adhesion molecule-1, ICAM-1) monoclonal antibody. Arthritis Rheum 35:S43–S43 (abstr)
Kavanaugh A, Norris S, Rothlein R, Nichols LA, Lipsky PE 1992b Pharmacokinetic analysis of rheumatoid arthritis patients treated with an anti-CD54 (intercellular adhesion molecule-1, ICAM-1) monoclonal antibody. Arthritis Rheum 35:S106–S106 (abstr)
Lipsky PE, Rosenthal AS 1975 Macrophage–lymphocyte interaction. II. Antigen-mediated physical interactions between immune guinea-pig lymph node lymphocytes and syngeneic macrophages. J Exp Med 141:138–154
Makgoba MW, Sanders ME, Luce GEG et al 1989 Intercellular adhesion molecule-1 (ICAM-1) monoclonal antibody inhibits cytotoxic T lymphocyte recognition. Ann NY Acad Sci 532:427–428
Matsui M, Temponi M, Ferrone S 1987 Characterization of a monoclonal antibody-defined human melanoma-associated antigen susceptible to induction by immune interferon. J Immunol 139:2088–2095
Mentzer SJ, Rothlein R, Springer TA, Faller DV 1988 Intercellular adhesion molecule-1 (ICAM-1) is involved in the cytolytic T lymphocyte interaction with human synovial cells. J Cell Physiol 137:173–178
Mileski WJ, Borgstrom D, Lightfoot E et al 1992 Inhibition of leukocyte–endothelial adherence following thermal injury. J Surg Res 52:334–339
Mileski WJ, Sikes P, Atiles L, Lightfoot E, Lipsky P, Baxter C 1993 Inhibition of leukocyte adherence and susceptibility to infection. J Surg Res 54:349–354
Natali P, Nicotra MR, Cavaliere R et al 1990 Differential expression of intercellular adhesion molecule 1 in primary and metastatic melanoma lesions. Cancer Res 50:1271–1278

Perry MA, Granger DN 1991 Role of CD11/CD18 in shear rate-dependent leukocyte–endothelial cell interactions in cat mesenteric venules. J Clin Invest 87:1798–1804
Pober JS, Gimbrone MA Jr, Lapierre LA et al 1986 Overlapping patterns of activation of human endothelial cells by interleukin 1, tumor necrosis factor and immune interferon. J Immunol 137:1893–1896
Raine CS, Lee SC, Scheinberg LC, Duijvestijn AM, Cross AH 1990 Adhesion molecules on endothelial cells in the central nervous system: an emerging area in the neuroimmunology of multiple sclerosis. Clin Immunol Immunopathol 57:173–187
Simmons D, Makgoba MW, Seed B 1988 ICAM, an adhesion ligand of LFA-1, is homologous to the neural cell adhesion molecule NCAM. Nature 331:624–627
Smith CW, Rothlein R, Hughes BJ et al 1988 Recognition of an endothelial determinant for CD18-dependent neutrophil adherence and transendothelial migration. J Clin Invest 82:1746–1756
Smith CW, Marlin SD, Rothlein R, Toman C, Anderson DC 1988 Cooperative interactions of LFA-1 and Mac-1 with intercellular adhesion molecule-1 in facilitating adherence and transendothelial migration of human neutrophils in vitro. J Clin Invest 83:2008–2017
Sobel RA, Mitchell ME, Fondren G 1990 Intercellular adhesion molecule-1 (ICAM-1) in cellular immune reactions in the human central nervous system. Am J Pathol 136:1309–1316
Staunton DE, Marlin SD, Stratowa C, Dustin ML, Springer TA 1988 Primary structure of intercellular adhesion molecule 1 (ICAM-1) demonstrates interaction between members of the immunoglobulin and integrin supergene families. Cell 52:925–933
Staunton DE, Dustin ML, Erickson HP, Springer TA 1990 The arrangement of the immunoglobulin-like domains of ICAM-1 and the binding sites for LFA-1 and rhinovirus. Cell 61:243–254
Van Seventer GA, Shimizu Y, Horgan KJ, Luce GEG, Webb D, Shaw S 1991 Remote T cell co-stimulation via LFA-1/ICAM-1 and CD2/LFA-3: demonstration with immobilized ligand/mAb and implication in monocyte-mediated co-stimulation. Eur J Immunol 21:1711–1718
Vejlsgaard GL, Ralfkiaer E, Avnstorp C, Czajkowski M, Marlin SD, Rothlein R 1989 Kinetics and characterization of intercellular adhesion molecule-1 (ICAM-1) expression on keratinocytes in various inflammatory skin lesions and malignant cutaneous lymphomas. J Am Acad Dermatol 20:782–790
Wacholtz MC, Patel SS, Lipsky PE 1989 Leukocyte function-associated antigen 1 is an activation molecule for human T cells. J Exp Med 170:431–448
Wegner CD, Gundel RH, Reilley P, Haynes N, Letts LG, Rothlein R 1990 Intercellular adhesion molecule-1 (ICAM-1) in the pathogenesis of asthma. Science 247:456–459

DISCUSSION

Barker: I'm surprised that you chose rheumatoid arthritis to study as an autoimmune disease because it is such a difficult disease to assess.

Rothlein: If you wanted to look at an autoimmune disease with an easy read-out, which would you pick?

Barker: What about asthma?

Rothlein: What would be an end-point? Would you treat somebody before they had an attack and see how long you could prolong their freedom from attack?

Barker: One thing you could do, if you were just looking at a model of autoimmunity, would be to look at aspirin-sensitive asthma. You could give these patients your antibody and then challenge them with aspirin to see whether they develop their asthma.

Rothlein: But there is the ethical issue of sensitizing somebody to a mouse antibody for the sake of an experiment. At least the rheumatoid arthritis patients have the potential of benefiting.

Of all the autoimmune diseases one looks at, rheumatoid arthritis has a fairly easy read-out, even though it is subjective. At least you can see something within weeks.

Haskard: I do not agree with that. The important read-out is disability, which is exceptionally difficult to predict. We already have ways of ablating the acute-phase response without necessarily influencing the long-term outcome.

Rothlein: I'm not arguing with the assertion that rheumatoid arthritis is difficult for obtaining approvable endpoints for the regulatory agencies. I'm saying that, as far as proof of concept is concerned, rheumatoid arthritis is a good model of an autoimmune disease.

Riethmüller: I think you are making life hard for yourself by using patients who have an average disease duration of about 18 years. All of us who have been involved in rheumatoid arthritis work with patients who already have obstruction or destruction of joints: to see a dramatic improvement in these patients is to expect a miracle and is asking too much of any drug.

The point is, can you alter the course of the disease in its early stages?

Rothlein: That is something that we will look at, probably with a 'humanized' version of the antibody.

Stanley: Are most of the antibodies against this mouse antibody you are giving anti-idiotypic?

Rothlein: Yes.

Stanley: So will humanizing the antibody help that situation?

Rothlein: That's a good question—the antibody sitting on an antigen-presenting cell will presumably inhibit, at least partially, the antigen presentation. The other side is: are we making anti-idiotypic antibodies because we have the mouse antibodies acting as an immunogenic carrier? I'm more concerned with an anti-isotype than an anti-idiotype, because an anti-idiotype will prevent the anti-ICAM-1 antibody from binding to ICAM-1 so you are not going to deposit immune complexes on the vasculature. An anti-isotype will. However, where we've seen immune complexes on the vasculature, we haven't seen this causing any problems in animals.

Wagner: Why don't you consider targeting P-selectin in rheumatoid arthritis? This molecule has been shown to be expressed in the affected joints.

Rothlein: At the time we started, P-selectin had yet to be identified as an adhesion molecule. It takes an incredible amount of work to get an antibody ready for the clinic.

Etzioni: Have you seen any side effects with your antibody treatments?

Rothlein: Some of the rheumatoid arthritis patients got headache, fever and chills about 2–7 h after the administration of the first dose of the antibody. As a result of this, we looked to see if ICAM-1, by being cross-linked, is able to transmit a signal (Rothlein et al 1994). Indeed, it turned out that cross-linked ICAM-1 induces a co-stimulated oxidative burst from CD14-expressing monocytes, suggesting that it is capable of delivering a transmembrane signal.

Etzioni: I have a question regarding your protocol of five days versus two days, which is similar to that which we used in an intravenous γ-globulin treatment for autoimmune disorder. There it was shown that in most patients if you give the two day high dose it will be as effective as a five day dose. Do you have any idea why, in your two day protocol when you gave a high dose (which should block the same number of receptors), you saw a different effect?

Rothlein: My speculation is that one has to interfere with a cascade of inflammatory events in order to break the cycle involved in rheumatoid arthritis. Perhaps breaking the cycle for two days isn't long enough, whereas five days is.

Elices: What is the half-life of the murine antibody?

Rothlein: About 1.5 days.

Elices: You mentioned that in the rheumatoid arthritis trial you didn't sec a delayed-type hypersensitivity response during treatment, however, you saw lymphocytosis. This would suggest to me that you are interfering with recirculation and migration patterns. Are you concerned about immunosuppressive effects?

Rothlein: During the treatment there has been no sign of adverse immunosuppressive effects. The transplantations were done in the presence of cyclosporin A and Imuran, and the patients have not come down with any opportunistic infections at any greater frequency than historical non-anti-ICAM-1 treated patients. Again, if you look at the LAD patients, viral infections seem to be the least of their problems.

Pals: The long-standing responses in rheumatoid arthritis suggest that you are somehow interfering with a vicious circle. This can be done just as effectively by other methods: for instance, favourable results have been reported with anti-TNF.

Haskard: What we need for the treatment of rheumatoid arthritis are not just drugs that affect the acute inflammatory process but drugs that will ultimately prevent disability. It's a bit disconcerting that anti-ICAM-1 has no effect on the acute-phase response, because one laboratory marker of long-term outcome in rheumatoid arthritis is a maintained high C-reactive protein level (CRP). Obviously, you're not in a position to tell us whether the antibody could have any effect on long-term disability, but I would be interested in your thoughts on why there's no effect on the CRP.

Rothlein: One has to figure that if it's blocking the trafficking stage there's still going to be inflammation going on in the joint for a period. We will have to see what would happen with the longer-term therapy with non-immunogenic anti-ICAM-1.

Haskard: The synthesis of CRP is being driven by blood cytokines. Have you looked at levels of circulating IL-6 or TNF?

Rothlein: Peter Lipsky is looking at TNF now. We have been looking at levels of circulating ICAM-1 and they don't seem to vary. One can take that as a marker of general inflammation.

Stanley: It as been pointed out here very well that if you look at integrin–ligand interactions you can take pieces of the ligand and interfere with those integrin–ligand interactions. If patients already have circulating ICAM-1, why doesn't that interfere with the ICAM-1 ligand–integrin interaction?

Rothlein: It's very difficult to get the circulating soluble ICAM-1 to bind to $\alpha_L\beta_2$ that's a monomer on the cell surface. The affinity of soluble ICAM-1 for $\alpha_L\beta_2$ is very low. There have been several reports showing that you can block certain aspects of T cell function with soluble ICAM-1, but again levels go much higher than one finds in circulation.

Verrando: Would this treatment be useful for multiple sclerosis (MS)?

Rothlein: A lot of people are interested in looking at that. It is clear that ICAM-1 is expressed in the astrocytes in MS plaques. There's no reason to think that if it works for rheumatoid arthritis it wouldn't also work in the acute phases of MS.

Pober: The attention is obviously directed to what's going on with leukocytes and lymphocytes. Platelets also express $\alpha_L\beta_2$ and could interact with ICAM-1; in your stroke model, were there any effects on coagulation parameters?

Rothlein: Yes, we've looked at other patients for effects on platelet function, and the answer is no. The platelet interactions with monocytes seem to be only P-selectin mediated. If there is $\alpha_L\beta_2$ on the platelets, it is not dependent on ICAM-1 for function.

Pober: I don't know that $\alpha_L\beta_2$ on platelets has an important role, but I just wonder whether ICAM-1 could serve other functions than leukocyte adhesion.

Rothlein: Yes, I would be more concerned with that question on the basis of the recent report that ICAM also binds fibrinogen (Languino et al 1993).

Winn: On the stroke model in the rabbit, the injury looks like it is caused by ischaemia and not reperfusion.

Rothlein: I would argue that when one injects microthrombi one doesn't get total ischaemia; I think ischaemia and reperfusion are happening at the same time and there is not a total blockage.

References

Languino LR, Plescia J, Duperray A et al 1993 Fibrinogen mediates leukocyte adhesion to vascular endothelium through an ICAM-1-dependent pathway. Cell 73:1423–1434

Rothlein R, Kishimoto TK, Mainolfi E 1994 Cross-linking of ICAM-1 induces co-signalling of an oxidative burst from mononuclear leukocytes. J Immunol 152:2488–2495

Identification of endogenous protein-associated carbohydrate ligands for E-selectin

Thakor P. Patel*, Christopher J. Edge*, Raj B. Parekh*, Susan E. Goelz† and Roy R. Lobb†

**Oxford GlycoSystems Ltd, Hitching Court, Blacklands Way, Abingdon, Oxon OX14 1RG, UK and †Biogen Inc, 14 Cambridge Center, Cambridge, Massachusetts 02142, USA*

Abstract. A comparative analysis of carbohydrate libraries derived from cell lines binding E-selectin was used to identify endogenous protein-associated carbohydrate ligands for E-selectin. Three structures, which together constitute less than 1% of the total cell surface protein-associated carbohydrate, were unique to cell lines capable of binding E-selectin, including neutrophils and the monocytic cell line U937. All are tetra-antennary *N*-linked structures with a sialic acid $\alpha 2 \rightarrow 3$ galactose $\beta 1 \rightarrow 4$ (fucose $\alpha 1 \rightarrow 3$) *N*-acetyl glucosamine $\beta 1 \rightarrow 3$ galactose $\beta 1 \rightarrow 4$ (fucose $\alpha 1 \rightarrow 3$) *N*-acetyl glucosamine lactosaminoglycan extension (sialyl-di-Lewis X [S-diLe^x]) on the arm linked through the C4 residue on the mannose. While all contained the expected 3-SLe^x sub-structure, these native structures have an additional fucosylated lactosamine unit. Direct evidence that these S-di-Le^x-containing structures are high-affinity ligands for E-selectin came from the use of recombinant soluble E-selectin–agarose affinity chromatography. These three carbohydrate structures bound specifically to the E-selectin column, while 3-SLe^x itself does not bind under identical conditions.

1995 Cell adhesion and human disease. Wiley, Chichester (Ciba Foundation Symposium 189) p 212–226

The specific adhesion of cells through the spatiotemporally controlled expression of ligands and their corresponding receptors is central to the development of multicellular organisms and to many normal processes within such organisms. One such process, which has lately been the focus of particular interest, is the adhesion and subsequent invasion into sites of inflammation of leukocytes, a process which is essential for effective defence of the host against infection and injury.

On the basis of various *in vivo* and *in vitro* studies, including those using intravital microscopy, a model is emerging in which leukocyte extravasation occurs in the following way. Leukocytes in the microcirculation are constantly coming into contact with the endothelial cell lining and are swept off this lining by the normal shear flow, unless the endothelial cells of the post-capillary venules (in the systemic microcirculation) express specific receptor/ligand pairs for leukocyte adhesion. In this case, the leukocytes roll with decreasing speed along the endothelium until they eventually come to rest, flatten against the endothelium and migrate through it. Various cell adhesion molecules are thought to be involved at different stages of this process and it is the successful involvement of different receptor/ligand pairs acting in sequence that ultimately leads to leukocyte extravasation. In broad outline, the selectins are involved in the initial adhesion between circulating leukocytes and inflamed endothelium, and serve to 'capture' leukocytes from the circulation and initiate the rolling process (Spertini et al 1992). Selectin-mediated rolling alone does not seem to be sufficient to bring leukocytes to rest. A second set of interactions involving one or more of the β_2 integrins is necessary to strengthen the adhesion of the leukocyte sufficiently to 'flatten' it against the endothelium and to prevent it being washed off (Lawrence & Springer 1991).

Three members of the selectin family of cell adhesion molecules have been reported to date, namely L-, P- and E-selectin (Bevilacqua & Nelson 1993). All three are glycoproteins with a similar and distinctive molecular architecture, including a calcium-dependent carbohydrate recognition domain at their N-termini. E-selectin is an inducible protein expressed on the surface of endothelial cells *in vitro* in response to cytokines such as interleukin (IL)-1 and tumour necrosis factor (TNF), and *in vivo* at inflammatory sites (Bevilacqua et al 1987). Using *in vitro* assays, cells have been identified which can bind to E-selectin, including peripheral blood cells such as neutrophils, monocytes, eosinophils, natural killer cells and myeloid-derived cell lines such as U937, HL-60 and THP1 (Bevilacqua et al 1989).

There is substantial evidence that specific carbohydrate structures can function as ligands for selectins through their lectin domains (Bevilacqua & Nelson 1993, Lasky 1992). In general, these ligands are acidic derivatives of the Lewis X and Lewis A trisaccharide structures, in which sialic acid or inorganic sulphate are attached to the C-3 position of the galactose residue. It has been shown that the surface expression of these structures is regulated by key fucosyl transferases (Goelz et al 1990), whose expression in cells that do not normally bind E-selectin can allow their subsequent adhesion to E-selectin-expressing cells via new sialylated fucose-containing structures on the cell surface. However, the acidic carbohydrates (such as 3-sialyl Lewis X) defined to date block adhesion *in vitro* with low affinity and with apparent K_ds usually of the order of 0.1–2 mM, depending on assay format (Brandley et al 1993).

Given the low affinities of the acidic carbohydrates so far defined for E-selectin, there remains uncertainty as to whether these acidic carbohydrates alone constitute the set of natural ligands for the selectins. We have therefore directed our efforts towards identifying naturally occurring carbohydrate ligands for

E-selectin, with a view ultimately to using the structure(s) of these natural ligands as the basis from which to design synthetic and non-carbohydrate antagonists of E-selectin-mediated adhesion.

Identification and characterization of natural protein-associated carbohydrate ligands for E-selectin

Analytical strategy

Carbohydrate libraries containing both *N*- and *O*-linked glycans were prepared from the pool of total plasma membrane glycoproteins isolated from cells that normally bind E-selectin avidly (polymorphonuclear leukocytes, U937 cells), which bind to E-selectin due to the expression of an exogenous added fucosyltransferase (COS cells transiently expressing elam-ligand fucosyl transferase, Goelz et al 1994) and ones that do not bind E-selectin at all (RAMOS cells and untransfected COS-7 cells). Each carbohydrate library was fractionated chromatographically and the primary structure was determined of those carbohydrates that were conserved, as judged by chromatographic identity, amongst the cell lines that bind E-selectin avidly. These primary structures were then compared to those of carbohydrates affinity-purified from the individual carbohydrate libraries using a recombinant, soluble E-selectin-agarose column.

Preparation and fractionation of carbohydrate libraries. The plasma membrane glycoprotein fraction was extracted from each cell line of interest (see above), and the associated carbohydrates were released, isolated and radiolabelled (by alkaline reduction) as described previously (Patel et al 1993). An aliquot of each pool of carbohydrate alditols was fractionated by QAE (quaternary amino ethyl) anion-exchange chromatography, high-performance anion-exchange chromatography (HPAEC) with pulsed amperometric detection (PAD) (using a Dionex Bio LC) and gel filtration (after de-sialylation), as also described previously (Parekh et al 1989).

The gel filtration chromatograms of the de-sialylated oligosaccharide alditol pools from each cell line are shown in Fig. 1. In each chromatogram, the single major oligosaccharide alditol (0–2) corresponds to galactose $\beta 1 \rightarrow 3$ *N*-acetylgalactosaminitol.

The regions of each chromatogram eluting between V_0 and 8 glucose units (gu.) are shown enlarged in Figs 1A′–1E′ (inclusive). Examination of these chromatograms reveals that, as judged by hydrodynamic volume, many oligosaccharide alditols are common to all the cell lines (though each occurs at a particular molar incidence in each pool), while some are not. Table 1 summarizes the data for those fractions which are not common to all cell lines. From Table 1, it is clear that asialo oligosaccharide alditols of hydrodynamic volume 24.4 gu. (U4), 27.3 gu. (U6), and 29.5 gu. (U8) are found only in the oligosaccharide pools from cells that bind E-selectin (polymorphonuclear leukocytes, and U937 and COS-7.2 cell lines). These asialo oligosaccharide alditols were therefore recovered for further structural analysis from the asialo oligosaccharide pool derived from U937 cells.

Primary structural analysis of asialo oligosaccharide alditols U4, U6 and U8 was performed using a combination of sequential exoglycosidase digestion, controlled acetolysis (previously optimized to maximize the selective cleavage of the mannose $\alpha 1 \rightarrow 6$ bond), and matrix-assisted laser desorption mass spectrometry to determine the molecular mass of the underivatized form of each alditol. During sequential exoglycosidase digestion and controlled acetolysis, P4 gel filtration chromatography was used to determine the change in hydrodynamic volume induced by each exoglycosidase or by the acetolysis procedure. From the induced change in hydrodynamic volume, the number of monosaccharide residues cleaved is readily calculated.

No simple 3-SLeX sub-structure was found in U4, U6 or U8, and all outer arm fucose $\alpha 1 \rightarrow 3$ residues were on the GlcNAc $\beta 1 \rightarrow 4$ Man $\alpha 1 \rightarrow 3$ branch and only in di-Lewis X structures. The assignment of the lactosaminoglycan extension to the GlcNAc $\beta 1 \rightarrow 4$ Man $\alpha 1 \rightarrow 3$ is based primarily on data from sequential exoglycosidase digestion and controlled acetolysis. Insufficient material was available to confirm the assignment by ^{1}H-nuclear magnetic resonance. The sialylation pattern of each of U4, U6 and U8 was determined by isolating the sialylated form of each structure from the original pool using a combination of QAE-anion exchange chromatography and HPAEC with PAD. Sialic acid linkages were assigned using the neuraminidases from *Arthrobacter ureafaciens* (which cleaves both $\alpha 2 \rightarrow 3$ and $\alpha 2 \rightarrow 6$-linked sialic acids) and from Newcastle disease virus (which cleaves principally $\alpha 2 \rightarrow 3$-linked sialic acid). Aliquots of the sialylated form of each of U4, U6 and U8 were individually treated with these neuraminidases before or after treatment with endo-β-galactosidase. The final structural assignments for the three carbohydrates associated with cell lines that are capable of binding E-selectin are shown in Fig. 2. These structures present not only on U937 cells, but also on human neutrophils, share a common S-di-Le^x structure located only on the C4 mannosyl arm of a tetra-antennary *N*-linked carbohydrate structure.

Fractionation of the carbohydrate alditol pool recovered from U937 cell plasma membranes using recombinant soluble E-selectin agarose. To provide direct evidence that the three structures identified from a comparison of carbohydrate libraries (Fig. 2) can bind to E-selectin, and so might be naturally occurring E-selectin ligands, we made an affinity column with recombinant soluble E-selectin. An aliquot (about 2 nanomoles at a loading concentration of about 0.2 μM) of the total oligosaccharide alditol pool from the U937 cell plasma membrane was passaged through a column of recombinant soluble E-selectin agarose. Unbound and weakly retarded material was washed through in loading buffer, and bound

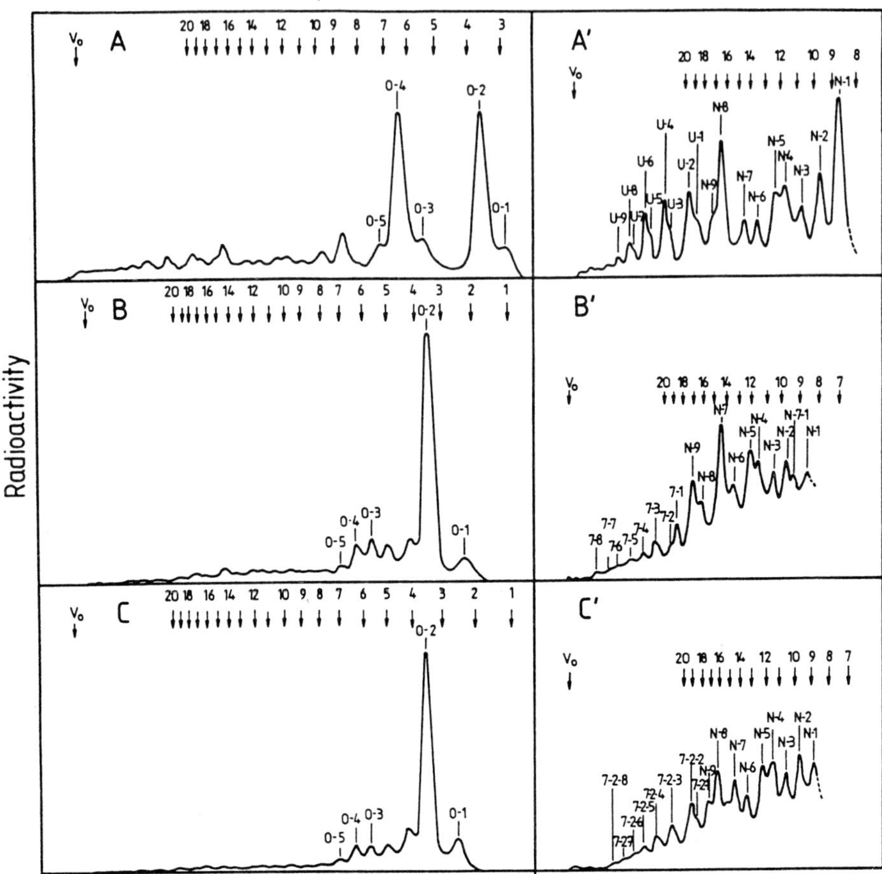

FIG. 1. Bio-Gel P-4 (~400 mesh) gel filtration chromatograms of the asialo carbohydrate alditols recovered from the plasma membranes of U937 (A), COS-7 (B), COS-7.2 (C), polymorphonuclear leukocytes (D), and RAMOS (E) cells. The corresponding regions of the chromatograms between V_0 and the glucose octamer (eluting at glucose unit = 8) are shown expanded in Figs 1A′, 1B′, 1C′, 1D′ and 1E′, respectively.

alditols were then eluted using buffered EDTA. The elution chromatogram is shown in Figure 3A′. The retained oligosaccharide alditol fraction (about 0.7% of the total) was de-salted, and this fraction was analysed as follows.

An aliquot of the bound material from the U937 cells was de-sialylated by incubation with neuraminidase (ex. *Arthrobacter ureafaciens*) and the de-sialylated pool fractionated by P4 gel filtration chromatography (Fig. 3B). Three alditol peaks are detected of elution volume 24.5 gu., 27.2 gu. and 29.6 gu. The identity of these peaks with U4, U6 and U8, respectively was confirmed by sequential exoglycosidase digestion. The remainder of the pool of bound material was fractionated into its (three) main components by HPAEC-PAD, and the

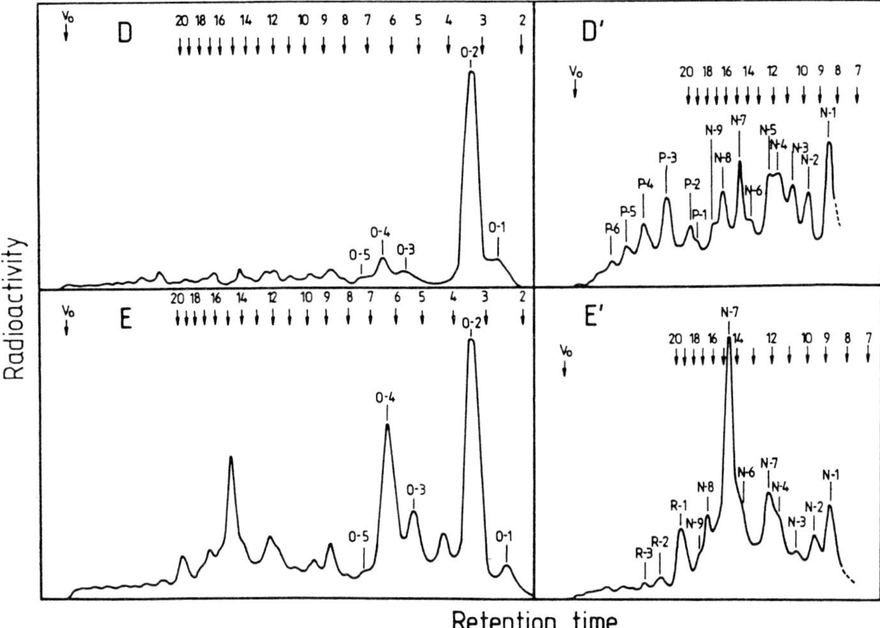

FIG. 1B.

sialic acid linkage to the arm carrying di-Lewis X was determined for each fraction by analysing Lewis X-bearing monosialylated fragments released using endo-β-galactosidase. It was found that the di-Lewis X extension was capped only with NeuAc $\alpha 2 \rightarrow 3$ in the three main components of the bound pool. From these results, we conclude that the major oligosaccharide structures released

TABLE 1 Occurrence of individual asialo oligosaccharides not common to all cell lines investigated

Hydrodynamic volume (gu.)	*Cell line*				
	U937	*COS-7.2*	*PMN*	*COS-7*	*RAMOS*
23.7	+	−	−	+	+
24.4	+	+	+	−	−
26.4	+	−	−	+	−
27.3	+	+	+	−	−
28.7	+	−	−	+	−
29.5	+	+	+	−	−

+, present; −, absent; gu., glucose units.

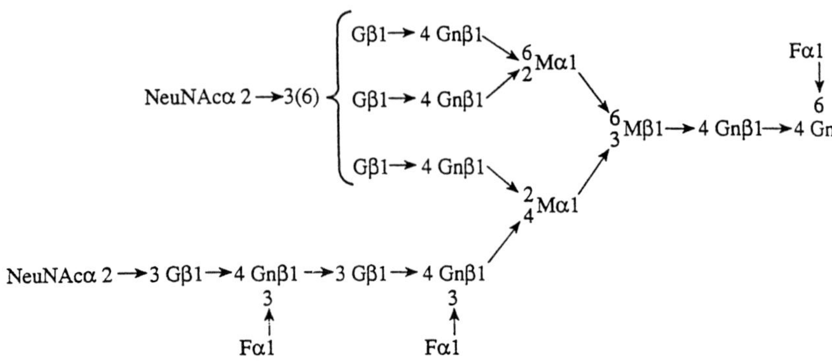

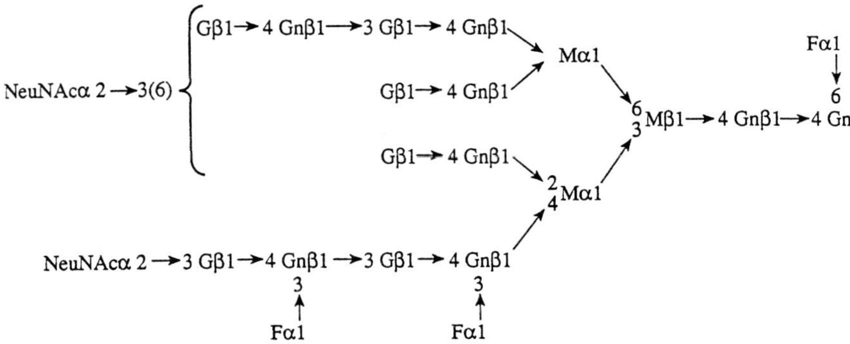

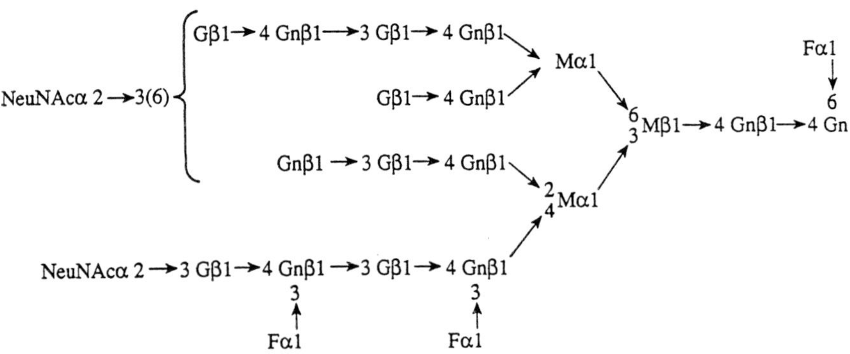

FIG. 2. A summary of the oligosaccharide structures proposed to be ligands for E-selectin. Assignation of these structures as ligands for E-selectin is based on the association of these structures with cell lines that are able to bind E-selectin and on the binding of these structures to a column of recombinant soluble E-selectin agarose. NeuNAc, *N*-acetylneuraminic acid; G, D-galactose; Gn, *N*-acetyl-D-glucosamine; M, D-mannose; F, L-fucose.

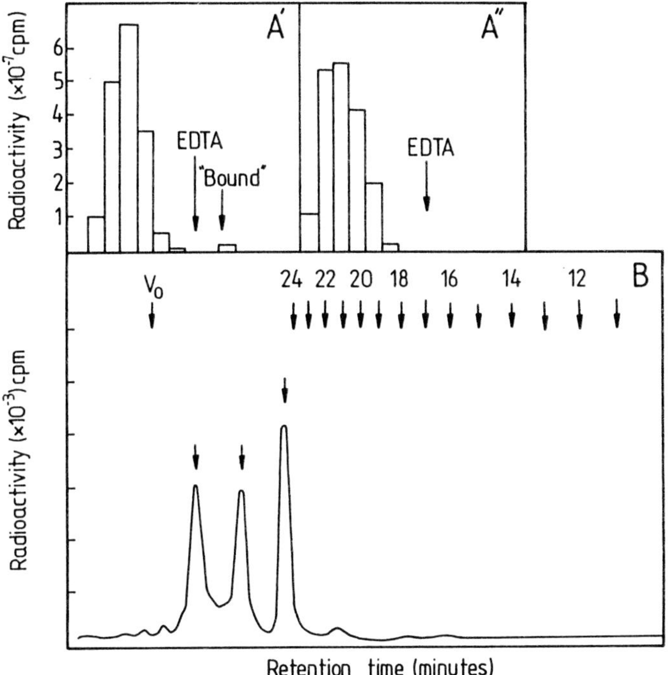

FIG. 3. Fractionation (A′) and analysis (B) of the oligosaccharide alditols from the plasma membrane of U937 cells after affinity chromatography using recombinant soluble (rs) E-selectin agarose. An aliquot of the oligosaccharide alditol pool was passed through an affinity column of rsE-selectin (A′). Bound alditols were eluted using EDTA. Authentic 3-SLe^x and 3-SLe^a were not retained by this affinity column (A″). An aliquot of the bound alditols was de-sialylated and fractionated by gel filtration (B). Arrowed peaks correspond in primary structure to the de-sialylated forms of the alditols shown in Fig. 2 as shown by correspondence in hydrodynamic volume and analysis by sequential exoglycosidase digestion.

from the U937 cell plasma membranes that are bound by recombinant soluble E-selectin agarose correspond to those assigned (Fig. 2) by the comparison of carbohydrate libraries, in which the lactosamine-bearing arm is capped with NeuNAc $\alpha 2 \rightarrow 3$. Authentic 3-SLe^x and 3-SLe^a were also passed through the E-selectin affinity column and, under the chromatographic conditions used for the U937 membranes, neither showed any binding (Fig. 3A″) indicating that U4, U6 and U8 bind E-selectin with higher affinity than either 3-SLe^x and 3-SLe^a. As a further control, the U937 oligosaccharide pool was also passed through a column to which bovine serum albumin had been coupled. No detectable material was bound (results not shown).

Fractionation of the carbohydrate alditol pools recovered from plasma membranes of U937, COS-7, COS-7.2 and polymorphonuclear (PMN) cells using recombinant soluble E-selectin agarose. An aliquot of the total oligosaccharide alditol pool from the plasma membranes of PMN, COS-7.2 and COS-7 cells were fractionated using recombinant soluble E-selectin agarose as described above for the oligosaccharide alditol pool from U937 cells. The relative percentages of bound oligosaccharide alditols were 0.6%, 0.2% and non-detectable (<0.01%), respectively. Bound oligosaccharide alditol fractions in the pools from COS-7.2 and PMN cells were analysed as described above for the bound material derived from U937 cells; the results indicate that the oligosaccharide alditols bound to recombinant soluble E-selectin in both the COS-7.2- and PMN-derived carbohydrate pools correspond to those shown in Fig. 2. Thus the major oligosaccharides on the plasma membrane of U937, COS-7.2 and PMN cells that are bound most strongly to recombinant soluble E-selectin are the three structures indicated in Fig. 2. These results suggest that PMN, U937 and COS-7 cells can synthesize the same basic complex carbohydrate structures, and that PMN and U937 possess a naturally expressed elam-ligand fucosyltransferase-like activity.

Discussion

The two complementary strategies used here to identify endogenous high-affinity protein-associated carbohydrate ligands for E-selectin lead us to the following conclusions:

(1) three related carbohydrate structures (Fig. 2) constitute the set of endogenous high-affinity protein-associated carbohydrate ligands for E-selectin;
(2) all possess an unusual 3-sialyl di-Lewis X extension on one arm of an *N*-linked tetra-antennary glycan;
(3) of the cells tested, all are present only on E-selectin-binding leukocytes and leukocytic cell lines;
(4) all bind to E-selectin with a relatively high affinity (estimated $K_d < 1\ \mu M$), greater than that of 3-SLe^x or 3-SLe^a;
(5) these structures represent a very small percentage of the total plasma membrane protein-associated carbohydrate.

The three identified carbohydrates have a high affinity for E-selectin as shown by the observation that the recombinant soluble E-selectin agarose column retained these structures (but not 3-SLe^x or 3-SLe^a) under the chromatographic conditions employed. These three carbohydrates were loaded onto the affinity column at a concentration of approximately 2 μM (1% of 0.2 mM) and, on the basis of the column size and flow-rate, a crude estimate for the on-column K_d between these carbohydrates and E-selectin is $<1\,\mu M$. If the significant difference in affinity between the carbohydrate structures reported here and 3-SLe^x and 3-SLe^a holds *in vivo*, it is likely that the carbohydrates identified here would be physiologically relevant ligands for E-selectin.

Sulphated carbohydrate ligands for E-selectin have been isolated from an ovarian cystadenoma protein (Yuen et al 1992). Our data suggest that sulphated carbohydrates are not endogenous protein-associated high-affinity ligands for E-selectin, at least on myeloid cells, as sulphated monosaccharides would have been recovered intact during the analytical process adopted here. This is consistent with most data on selectin–carbohydrate interactions, which suggest that sulphated carbohydrates are important primarily in the interaction of L-selectin with high endothelial venules in peripheral nodes (Lasky 1992). However, the experimental approach adopted here was designed specifically to detect *N*-linked or *O*-linked protein-associated carbohydrate ligands, and would not have led to the discovery of glycolipid- or proteoglycan-associated ligands, which may be sulphated.

These data are relevant to the design of carbohydrates and carbohydrate mimetics that can selectively antagonize selectin-mediated adhesion. Although 3-SLe^x itself is a low-affinity ligand for selectins, 3-SLe^x-based carbohydrate structures of higher affinity have been synthesized, and extension of the 3-SLe^x structure at its reducing terminus with an aliphatic C8 carbon chain provides significantly increased affinity (Nelson et al 1993). Such structures may mimic the extended S-diLe^x structure we describe here, and suggest that selectins may have an extended groove or binding pocket capable of multiple interaction points. Analysis of the crystal structure of E-selectin reveals a potential binding domain for 3-SLe^x that could also accommodate S-diLe^x (Graves et al 1994). The use of the S-diLe^x structure may have significant advantages over SLe^x itself in examining E-selectin binding interactions, including use in crystallographic approaches and in designing higher-affinity, univalent antagonists.

References

Bevilacqua MP, Nelson RM 1993 Selectins. J Clin Invest 91:379–387

Bevilacqua MP, Pober JS, Mendrick DL, Cotran RS, Gimbrone MA 1987 Identification of an inducible endothelial–leukocyte adhesion molecule. Proc Natl Acad Sci USA 84:9238–9242

Bevilacqua MP, Stengelin S, Gimbrone MA Jr., Seed B 1989 Endothelial leukocyte adhesion molecule-1: an inducible receptor for neutrophils related to complement regulatory proteins and lectins. Science 243:1160–1165

Brandley BK, Kiso M, Abbas S et al 1993 Structure-function studies on selectin carbohydrate ligands. Modifications to fucose, sialic acid and sulphate as a sialic acid replacement. Glycobiology 3:633–639

Goelz S, Kumar R, Potvin B, Sundaram S, Brickelmaier M, Stanley P 1994 Differential expression of an E-selectin ligand [SLE(X)] by two Chinese hamster ovary cell lines

transfected with the same alpha(1,3)-fucosyltransferase gene (Elft). J Biol Chem 269:1033–1040
Graves BJ, Crowther RL, Chandran C et al 1994 Insight into E-selectin/ligand interaction from the crystal structure and mutagenesis of the LEC/EGF domains. Nature 367:532–538
Lasky LA 1992 Selectins: interpreters of cell-specific carbohydrate information during inflammation. Science 258:964–969
Lawrence MB, Springer TA 1991 Leukocytes roll on a selectin at physiologic flow rates: distinction from and prerequisite for adhesion through integrins. Cell 65:859–873
Nelson RM, Dolich S, Aruffo A, Cecconi O, Bevilacqua MP 1993 Higher-affinity oligosaccharide ligands for E-selectin. J Clin Invest 91:1157–1166
Parekh RB, Dwek RA, Thomas JR et al 1989 Cell-type-specific and site-specific N-glycosylation of type-I and type-II human-tissue plasminogen-activator. Biochemistry 28:7644–7662
Patel T, Bruce J, Merry A et al 1993 Use of hydrazine to release in intact and unreduced form both N-linked and O-linked oligosaccharides from glycoproteins. Biochemistry 32:679–693
Spertini O, Luscinskas FW, Gimbrone MA Jr, Tedder TF 1992 Monocyte attachment to activated human vascular endothelium *in vitro* is mediated by leukocyte adhesion molecule-1 (L-selectin) under nonstatic conditions. J Exp Med 175:1789–1792
Yuen CT, Lawson AM, Chai WG et al 1992 Novel sulfated ligands for the cell-adhesion molecule E-selectin revealed by the neoglycolipid technology among O-linked oligosaccharides on an ovarian cystadenoma glycoprotein. Biochemistry 31:9126–9131

DISCUSSION

Wagner: Do the carbohydrate structures that you have identified as binding to the E-selectin column (Fig. 2) also bind P-selectin?

Parekh: We don't know, but we would like to find out.

Wagner: How do the proteins that you have immunoprecipitated with your antibody from polymorphonuclear leukocytes compare to the E-selectin ligands identified by McEver, Vestweber, Jutila and others?

Parekh: We haven't done detailed comparisons. Whilst we don't yet have supporting data, we think that the proteins we precipitated may be related to the LAMP proteins, which are very heavily *N*-glycosylated with polylactosamine extensions.

Humphries: Alan Williams predicted that affinity isn't of major importance for carbohydrate ligands and that, instead, the on-rate is most important. Can you say something about the on and off rates that you have determined for your SLe^x binding?

Parekh: There's nothing unusual about them. The low affinity comes primarily from the high off-rate.

Humphries: It will be interesting to measure the rate constants for your authentic carbohydrates, to test whether their high activity is due to a truly high affinity.

Parekh: It has often been assumed that carbohydrate-mediated binding will only involve multiple low-affinity interactions which lead to high avidity. It is now clear that not every carbohydrate interaction can be generalized in this way.

Humphries: But isn't it also the conformational flexibility in a sugar that gives a high on-rate as well, not necessarily just a clustering effect?

Parekh: If one considers the molecular dynamics and conformational studies on carbohydrates, it is clear that they are far more rigid molecules than peptides of the same number of monomers. The conformational fluidity is the hallmark of short linear peptide polymers, but not of carbohydrate structures.

Stanley: The conformation of the E-selectin–Fc chimera must be important if you are looking at carbohydrate binding to it, so how do you make that molecule so that the conformation is correct?

Parekh: The molecule was expressed in CHO cells. No information has been obtained on the authenticity of its conformation, other than to show that its binding properties are as expected.

Riethmülller: We've been interested in Lewis Y: it is one of the most attractive targets for tumour-directed therapy. Have you looked at phage display libraries? All the antibodies we have are very low affinity, and they are all more or less generated independently of T cells. Can these libraries be used to generate antibodies with a high affinity?

Parekh: I don't know.

Pober: It's not clear that all high-affinity ligands are really involved in the mediation of adhesion. Eugene Butcher has argued, in his case for L-selectin being a relevant ligand for E-selectin, that neutrophils extend cellular processes and that relevant ligands for rolling need to be located on the tips of these processes in order for them to come near the endothelial cells (Picker et al 1991). Do you know anything about where these ligands are located on a resting neutrophil?

Parekh: No, the cellular histology needs to be done.

Wagner: About the specificity of the heptasaccharide component of the molecule shown in Fig. 2: does it inhibit all the selectins?

Parekh: Probably not L-selectin: the true ligand for L-selectin hasn't been confirmed yet, but there's a strong suggestion that it contains GalNAc sulphate and two fucoses. I think there's enough structural dissimilarity between that and this heptasaccharide for each to inhibit L- or E-selectin individually. P-selectin is the difficult one. My sense is that with the carbohydrate mimetic approach we are taking, we may end up with structures that are essentially incapable of distinguishing between E- and P-selectin.

Hynes: This might not be the case, because that's what people were worried about with the RGD mimetics. There are a number of integrins that are blocked by RGD and the initial concern was that you wouldn't be able to get one that would distinguish specific integrins. It turned out to be fairly easy to make peptides which could distinguish between different integrins.

Sanchez-Madrid: We have recent results showing that during activation B lymphocytes express SLe^x (Postigo et al 1994). This allows the activated B cell to bind to E- and P-selectins. The expression of SLe^x does not correlate with the concomitant expression of any of the fucosyl transferases so far described.

Riethmüller: When is the antigen expressed during B cell activation?

Sanchez-Madrid: It is quite early: you can see a good expression at 24 h, it peaks at 48 h and it declines at 72 h.

Shaltiel: Coming back to rigidity; is it clear that you need the rigidity, or just the distance? In other words, would it be possible to get a flexible molecule to bind the carboxyl and the three hydroxyls? Or do you think there is a mechanism of induced fit in which the carbohydrate affects the structure of the binding site in a way that imposes a better fit?

Parekh: I think the notion is that this would probably be involved, for the simple reason that one can't easily dock carbohydrates straight into the E-selectin crystal structure. From a pharmaceutical point of view, the more flexible the molecule, the more entropic energy is lost when it is constrained. So there's a balance between keeping it constrained enough that you don't lose excessive entropic energy and flexible enough to allow it to dock into the available binding site.

Ernst: When we connect the sialic acid and fucose with a flexible linker, we lose activity. Therefore we believe that the LacNAc core is essential for presenting the peripheral sugars in the correct spatial arrangement.

Shaltiel: With modulated structures, is it possible to get different cation dependencies?

Parekh: I don't know. The argument for Ca^{2+} in the carbohydrate recognition domain is that it chelates ring hydroxyl groups. I think the Ca^{2+} is quite critical.

Hynes: It is also not clear from the crystal structure of E-selectin what the EGF domain is doing, because the carbohydrate is over on the face of the lectin domain, away from the EGF domain, even though that domain is supposed to play a role. Do you have any theories about how this could be playing a role?

Parekh: I could come up with vague arm-waving arguments about it mediating the conformation of the carbohydrate recognition domain, but the mechanism is another matter.

Herrlich: With the sugar sticking out so far and the high specificity of the modifying enzyme, you could imagine that all the specificity of protein recognition is in the sugar. Is that the current view?

Parekh: If it was just a carbohydrate that directed the action of the fucosyl transferase, you would expect that essentially any carbohydrate structure which had that lactosamine would be a substrate for the fucosyl transferase. Yet even on that same molecule there are arms that could be fucosylated that aren't. Also, other proteins carry lactosamine *N*-glycan extensions. So my own model is that the proteins, to which these carbohydrate structures attach, direct the action of the fucosyl transferase. Therefore there may be a high degree of protein-determined glycosylation for these ligands.

Hynes: Can you estimate, from knowing how many of these carbohydrate groupings and how many proteins you have on the cell, how many carbohydrates of this type are on each of the proteins? Is it one group, or is this like a mucin with a lot of carbohydrates on one backbone?

Parekh: We think there are only two *N*-glycosylation sites; the way we're addressing this is to do a site analysis.

Ernst: Have you compared di-Lewis X and SLe^x in your ligand binding assay?

Parekh: No.

Ernst: You mentioned that the replacement of GlcNAc in SLe^x by 2,6-deoxy Glc does not affect the affinity in the competitive assay. Doesn't this contradict Michael Bevilacqua's results, which show a dramatic increase in affinity when GlcNAc in SLe^a is replaced by $GlcN_3$ or $GlcNH_2$?

Parekh: We didn't find that. We also converted the carboxyl of the sialic acid to an aldehyde and didn't find a significant improvement. There is still controversy over how carbohydrates bind to E-selectin because we don't have the binding in the crystal.

Elices: You talked about your strategy for drug development. What do you think of using the carbohydrate scaffold as the starting point for putting on the important residues? Do you think that you can use carbohydrate synthesis to do that or, alternatively, do you think you can get away from carbohydrates altogether?

Parekh: A monosaccharide is a very small intensely chiral centre with some degree of conformational flexibility based upon the oxygen in the ring and the hydroxyl in the exocyclic position. So as a molecular scaffold for permuting and presenting an enormous number of functionalities, a chemist almost couldn't have designed one better. You can decorate monosaccharide scaffolds with a variety of carbohydrate or non-carbohydrate substitutions, and the three dimensional space you can probe with them is very considerable.

Wagner: What is the half-life of these molecules, in comparison to peptides, in the bloodstream?

Parekh: No one has yet published their reports on the half-life of SLe^x, but the general view seems to be that the tetrasaccharide and its derivatives have a half-life in the order of tens of minutes.

Hynes: Are there ways you can modify your groupings to give them a longer half-life?

Parekh: We're trying to move to an organic mimetic. I don't know what the half-life of this will be, but that's one of the reasons that we didn't want to use SLe^x.

Riethmüller: What is the reason for that short half-life?

Parekh: It's thought to be removal by the kidney.

Ernst: Cytel knows best!

Elices: They're pretty stable, that's all I can say.

Shaltiel: Are there diseases caused by specific inhibitors of fucosyl transferase, or are there people defective in such enzymes?

Etzioni: In the Bombay blood phenotype, one of the fucosyl transferase enzymes is defective.

Riethmüller: Could you use synthetic colorigenic substrates to screen for such enzyme defects?

Parekh: Fucosyl transferase inhibitors have been proposed for a long time, because there are certain cancers where there's an elevation of an α-1,6 fucosyl transferase. The difficulty has been to establish clearly a causal link between elevated fucosylation and tumorigenicity.

Ernst: But these screens are normally done on isolated enzymes; in these assays the biggest obstacle to be overcome, namely cell permeability, is not addressed.

Wagner: Because these mimetic molecules can be small and simple, could they be given orally?

Parekh: Yes, that's the hope. That's why we are trying to get away from the carbohydrate. The glycosidic linkage involving sialic acid is pH labile, and the acidity in the stomach at 37 °C could substantially hydrolyse that linkage.

References

Picker LJ, Warnock RA, Burns AR, Doerschuk CM, Berg EL, Butcher EC 1991 The neutrophil selectin LECAM-1 presents carbohydrate ligands to the vascular selectins ELAM-1 and GMP140. Cell 66:921–933

Postigo AA, Marazuela M, Sánchez-Madrid F, de Landazuri MO 1994 B lymphocyte binding to P- and E-selectins is mediated through the *de novo* expression of carbohydrates on *in vitro* and *in vivo* activated human B cells. J Clin Invest 94, in press

Final discussion

Hynes: I started by posing some questions that we might answer at this symposium. One of these concerned which molecules to target, and how to do it. Another question concerned the differences between acute and chronic treatment. I raised the issue of gene therapy: I think the general answer for the diseases that have been discussed was that it is not particularly attractive, except perhaps for leukocyte adhesion deficiency (LAD) I.

We have a pretty good idea of the spectrum of adhesion molecules that exist, although I'm sure that there are still some that we don't know about. However, we don't know enough about where they occur (their localization) and which of them are expressed in different inflammatory situations. So we need more information from the clinicians about where adhesion molecules are expressed and which molecules are expressed in specific inflammatory diseases. With reference to interpreting human disease and animal models and applying the data from one to the other, we need some comparisons between the different systems—they're not always the same. One obvious way to do this is by immunohistochemistry; the more of these data that can be collected the better. It is probably worth thinking about some sort of database for this information. Databases are already being developed for the expression of different genes in development; it might be worthwhile to do this for human diseases as well. Dorian Haskard's imaging of adhesion molecule expression *in situ* in pigs suggests the possibility of doing this on human patients; that's a very interesting and appealing idea.

As a task for the animal model people, we need a lot more knockouts, because they provide information which is, to one degree or another, portable to the human system. They also block out some areas: in particular, you can identify adhesion molecules that are not important for some functions. In particular, it struck me during this meeting that we need knockouts of all the β_2 integrins separately and together. There isn't yet a full knockout of β_2 in the mouse—the one knockout that's been done so far is a hypomorph, although this is still an interesting mutant. In fact, even LAD I is a spectrum from null to hypomorph. We haven't talked much about subtle mutants, but that's clearly one of the things that's coming next. I only know of one further α integrin knockout in progress and I hope the others are going to get done.

During the discussion of the skin diseases, the issue came up of which of the desmosomal proteins is really important for each adhesion event. Knockouts would be really helpful here.

We now have a lot of single knockouts: all three selectins have been knocked out independently. It is probably going to be important to knock them out in pairs and even in triplets. The same is true of integrins, where it may be easier because they're not all linked, so it will be possible to intercross the knockouts. Experiments are in progress to make double-knockout selectin mutants. So the animal model people have a lot of work, the clinicians have a lot of work, and we have to try to put all of the results together.

In terms of adhesion molecule blockers, there's a real contrast between the antibody approach and the small molecule approach: everybody has their own preferences about this. Obviously, the main problem with antibodies is whether you can use them repeatedly without getting immune responses. Can one avoid this? How good is 'humanization'? Do you still get anti-idiotypic antibodies? We don't really know the answers to these questions, but perhaps someone has some thoughts on them. Certainly, antibodies are good for proof of concept.

For the peptide and the carbohydrate blockers, we really need some more crystal structures. It's clear from the case of the selectins that the crystal structure, while it doesn't answer all the questions, is very helpful. We desperately need an integrin crystal structure. None of the academics will do it, because they're afraid the companies are doing it. So if the companies are doing it, I hope they'll publish it, so that everyone can make use of the data. This is obviously not an easy project to do: the head of an integrin is large, but it is going to be hard to proceed beyond some of the issues that we have discussed without the crystal structure. I would have thought that the drug companies would want to do at least a β_2 integrin and GPIIb/IIIa.

The issue of clinical trials was discussed. They are difficult, they're expensive, and the endpoints are very difficult for some of these diseases. Clinical trials for cancer are particularly hard; without surrogate endpoints and some good ideas about how to refine these trials (e.g. by careful choice of target populations), it's going to be very hard to develop therapies. Even if we can figure out, let's say, that selectins are homing receptors for metastatic cells, doing trials on whether anti-selectin therapy does any good is going to be even more problematic than trials for rheumatoid arthritis.

Elices: I like the idea of having a databank for expression of adhesion molecules in human disease: it would be an invaluable resource for researchers in this field.

With regard to compounds developed for clinical trials, in my mind a big challenge that lies ahead of us is to show that anti-adhesion therapy really works in treating human disease. Potentially, there are two therapeutic approaches that may be tried. I favour the small molecule approach, principally because I'm interested in targeting chronic indications in which you administer a drug repeatedly for a long time.

Hynes: There's also the issue of oral delivery rather than injection. There's no way you're going to deliver an antibody orally.

Etzioni: We have to distinguish between two groups of patients: acutely and chronically ill. In acutely ill patients, such as the reperfusion/ischaemia patients

Bob Winn talked about, there is a place for antibody therapy, because this will have to be administered only once. But with a chronic condition, like rheumatoid arthritis, I think we will only be able to intervene with anti-adhesion therapy when the patient has just developed the disease, or even before. For example, we have some experimental results with NOD mice: as you know, the development of diabetes in NOD mice takes months, and in the mice who are still clinically well you can see the expression of adhesion molecules in the Langerhans' islets (Shehadeh et al 1994). So in those conditions, administering the anti-adhesion molecule at appropriate intervals may represent a clinical cure for the disease. But once a condition such as rheumatoid arthritis has been established for 10 years, it is less likely that those drugs will have a beneficial effect.

Winn: I agree with Amos Etzioni. There's a big difference between the acute and chronic patients. We know from the two patient groups (LAD I and LAD II) that chronic therapy will lead to increased risk of infection. Molecules for chronic therapy will have to be more specialized; I don't think that any molecules that are presently available are ready to be used in chronic therapy. In the acute cases, clinical trials could take place that would be wrapped up in a year or two.

Pober: In thinking of ways to inhibit these interactions you are focusing on molecules that are already expressed; one of the key points about the inflammatory process is that the resting endothelium is a negative surface as far as these interactions are concerned. Few of the relevant adhesion molecules are expressed. An alternative way to think about this problem is how to keep the endothelial surface as a negative one, or how to restore it to a negative state.

Hynes: That's a very good point. All the adhesion therapies have to be matched against, for instance, aspirin (if you're talking about thrombosis) or anti-inflammatory cytokines (if you're talking about inflammation). It may be much better to target the triggering molecule than the structural molecule that it triggers.

Winn: It may be that the triggering molecule is generated locally and therapeutic drugs cannot block its function before it activates the leukocytes or the endothelial cells. Therefore, beneficial effects may not be seen when the triggering molecule is blocked.

Labow: I would argue that for chronic disease you have chronically activated endothelial cells, so inhibiting endothelial activation would be very effective. However, that could actually be the most dangerous therapy, because it is potentially immunosuppressive.

Humphries: An alternative strategy would be to target the signalling molecules involved in adhesive processes. We've not really mentioned them at this meeting. We desperately need a 3D vision of how the cytoskeleton assembles an adhesion site and how this coordinates with signalling molecules. This may help us understand how specificity is generated.

Birchmeier: Concerning these alternative strategies, is it feasible to select drugs which interfere, for instance, with promoters?

Labow: Yes, a number of companies are doing this. Glaxo has a big project underway trying to identify activators of selectin expression. We have also done some work on that. Gene expression assays are very easy high flux screens to do. However, we would like to do those assays in a human endothelial cell line; to date, the big problem has been getting a really good differentiated human endothelial line that retains full cytokine inducibility. We don't have one available: the ones we have made by transforming HUVECs lose their cytokine inducibility after 10–15 passages. Other people are just taking primary lots of HUVECs and screening by ELISA for E-selectin expression.

Hynes: Do the endothelial experts have any thoughts on how to make a well-behaved endothelial cell line?

Wagner: Many people have tried.

Labow: There are some good mouse lines, so one possibility is to go into the mouse lines that seem to retain cytokine inducibility, although the basal level of expression in these lines has gone way up.

Wagner: We see this also with endothelial cells in culture. Without stimulating them at all, one cell out of 50 will be E-selectin positive.

Labow: We found that primary endothelial cells are extremely sensitive to triggering. If you shake a flask of endothelial cells you get expression all over the place! When we do transfection assays with them, we have found electroporation is one of the best methods for turning on all of the cytokine-inducible genes, so they are hard cells to work with.

Riethmüller: We have seen quite an interesting correlation between expression of adhesion molecules and clinical metastasis. In attempting to bring these results into the clinic, one should learn from the long and painful experiences we have had when we have attempted to use antibodies for cancer therapy. We have realized that the whole approach can fall into demise simply by targeting the wrong patient population. To try to show efficacy of antibodies in terminal cancer is love's labour lost. I would stress the point that the pace of progress in adjuvant therapy of cancer is so scandalously slow because we have no surrogate maker. The US Food and Drug Administration are learning their lesson with the long latency in HIV infection. There is also a real need for a surrogate maker which gives an endpoint that may not be the final answer but which would help to develop and to monitor therapies. Along those lines, I think that our technique of detecting minimal residual cancer might be an important technique for monitoring adjuvant therapy in solid tumours. At least 50% of our cancer patients are suffering from occult metastases when they first come to the clinic.

Parekh: What's the degree of confidence that antagonizing a single cell adhesion molecule–receptor interaction would abrogate an entire physiological process? Multiple cell adhesion molecules seem to be involved in an entire process of adhesion, so if your strategy is to knockout just one such group of molecules, won't there be some redundancy in the system? Therefore aren't we really talking about a combination therapy, where you have to know about sets of molecules which need to be deleted?

Hynes: I think you do need to know about sets, but the answer to that question depends very much on the individual case. In some adhesion events there is a redundancy and in others there are multiple groups involved but in series, so that you only need to block any one arm and you have blocked the whole adhesion process. Furthermore, you may not want to block adhesion entirely, because you may then disable the whole system and get infection. It might be better to tone adhesion down a little bit. That's certainly the case with the antithrombotic agents: you don't want to block platelet aggregation entirely because then you will bleed, but you want to block it about 50%.

Stanley: Recently, there has been a lot of progress on adhesion of basal cells (at least in the epidermis) to the underlying basement membrane. There have been antibody-mediated diseases as well as genetic diseases that have been well defined and in which the target molecules mediate adhesion. For example, antibodies against specific molecules like halinin, a laminin isoform, as well as genetic mutations in the same molecules, cause subepidermal blisters due to loss of basal cell adhesion (Domloge-Hultsch et al 1992). This is something we haven't mentioned.

Pober: I would like to comment, from the perspective of clinical transplantation, about the concerns that have been raised regarding what happens to patients if you effectively block these adhesion pathways. There is a margin between immunosuppression and immunodeficiency. The good news is that it's not a very narrow window at all. In fact, the transplanters have fine-tuned this for years so that you can achieve very effective immunosuppression without inducing immunodeficiency. In Bob Rothlein's presentation about the introduction of his antibodies into transplant recipients, they don't stop the other immunosuppressive drugs. These antibodies are used on top of existing therapy and the patients still have functional immune systems. Furthermore, what you're looking for in this setting is not a single magic bullet. Rather, you are looking for a new therapeutic that you might use as either a prophylactic or as an acute intervention for a rejection episode until you get back to the point where your basal regimen is effective again. The fear about antibodies to adhesion molecules producing total immunodeficiency has not been borne out by experience. The issue may be instead whether they are effective enough.

Ernst: Just a short remark on an alternative strategy for inhibiting adhesion: ISIS Pharmaceuticals has recently shown that the expression of endothelial adhesion molecules (ICAM-1, VCAM-1 and E-selectin) can be successfully inhibited with antisense oligonucleotides.

Hynes: To summarize in 15 seconds: I think the papers and discussions in this meeting have shown that there's obviously a lot of activity on the subject of cell adhesion and disease, but it is also clear that there are still enough questions for another meeting!

References

Domloge-Hultsch N, Gammon WR, Briggaman RA, Gil SG, Carter WG, Yancey KB 1992 Epiligrin, the major human keratinocyte integrin ligand, is a target in both an acquired autoimmune and an inherited subepidermal blistering skin disease. J Clin Invest 90:1628–1633

Shehadeh N, Karnieli A, Fidenboum G, Cahana F, Etzioni A 1994 Expression of ICAM-1 in pancreatic islets in NOD mice. In press

Index of contributors

Non-participating co-authors are indicated by asterisks. Entries in bold type indicate papers; other entries refer to discussion contributions.

Indexes compiled by Liza Weinkove.

Subject index